T0269250

Additive Manufacturing of Titanium Alloys

Additive Manufacturing of Titanium Alloys

State of the Art, Challenges, and Opportunities

Bhaskar Dutta

Francis H. Froes

AMSTERDAM • BOSTON • HEIDELBERG • LONDON
NEW YORK • OXFORD • PARIS • SAN DIEGO
SAN FRANCISCO • SINGAPORE • SYDNEY • TOKYO
Butterworth-Heinemann is an imprint of Elsevier

Butterworth-Heinemann is an imprint of Elsevier
The Boulevard, Langford Lane, Kidlington, Oxford OX5 1GB, UK
50 Hampshire Street, 5th Floor, Cambridge, MA 02139, USA

British Library Cataloguing-in-Publication Data
A catalogue record for this book is available from the British Library

Library of Congress Cataloging-in-Publication Data
A catalog record for this book is available from the Library of Congress

ISBN: 978-0-12-804782-8

For Information on all Butterworth-Heinemann publications
visit our website at https://www.elsevier.com

Publisher: Joe Hayton
Acquisition Editor: Christina Gifford
Editorial Project Manager: Heather Cain
Production Project Manager: Sruthi Satheesh

Typeset by MPS Limited, Chennai, India

CONTENTS

ABOUT THE AUTHORS

Dr. B. Dutta is chief operating officer at DM3D Technology, a leading additive manufacturing company. He received his Bachelor of Engineering from Calcutta University and his Master of Engineering as well as PhD in metallurgical engineering from Indian Institute of Science, Bangalore, India. He has 26 years of experience in the field of metallurgy and metal processing, including 11 years in the additive manufacturing (AM) industry. He has been directly involved in participating and directing AM research and technology development as well as commercial product development using AM. He has been the principal investigator and project lead for multiple federal government sponsored R&D projects involving development of AM technology. He has more than 50 publications, including about 15 publications in the field of AM and one chapter on AM in *Titanium Powder Metallurgy*. He also has eight pending patents in this area.

After receiving a BSc from Liverpool University and a MSc and PhD from Sheffield University all in physical metallurgy **Dr. F.H. (Sam) Froes** then became involved in the titanium field for more than 40 years. He was employed by a primary titanium producer, Crucible Steel Company, where he was leader of the Titanium group. He then spent time at the USAF Materials Lab where he was supervisor of the Light Metals group (which included titanium). This was followed by 17 years at the University of Idaho where he was an institute director and department head of the Materials Science and Engineering Department. He has over 800 publications, in excess of 60 patents, and has edited almost 30 books, the majority on various aspects of titanium. Recent publications include a comprehensive review of *Titanium Powder Metallurgy*, three articles on titanium additive manufacturing and books on all aspects of titanium science, technology and applications (ASM) and this book on titanium additive manufacturing. In recent years he has cosponsored four TMS Symposia on Cost Effective Titanium. He is a fellow of ASM, is a member of the Russian Academy of Science, and received the Service to Powder Metallurgy award by the Metal Powder Association and the Ben Gurion Medal from the Israeli Materials Society.

PREFACE

The past few years have shown significant advances in the additive manufacturing (AM) technologies leading to the production of fully functional parts using titanium and its alloys. While powder bed fusion technologies offer the ability to build hollow near-net shapes with finer resolution, directed energy based technologies offer the ability to add features on existing parts and remanufacture/repair damaged parts, as well as building parts directly from CAD data. Most of the studies in the industry reveal that the mechanical properties of AM material are as good as or better than the conventionally fabricated titanium alloys. Selection of the right AM technology, along with proper design optimization, can lead to very significant savings through greatly reduced buy-to-fly ratios, overall weight reduction, and scrap reduction. Additionally, these technologies offer design freedom that conventional manufacturing does not.

However, full exploitation of the benefits of AM depend largely on educating the manufacturing and design community and successful integration of these technologies in manufacturing industry and it is the aim of this book to help achieve these goals. The aerospace and medical industries have so far been the largest driver for the usage of titanium AM materials, while other industries, such as the automotive industry, are beginning to exploit benefits of AM of titanium alloys. The recent push in low cost titanium powders is expected to expand usage of AM in more cost sensitive industries such as automotive.

A book of this type could not be compiled without the help of many colleagues and the authors would like to recognize the contributions of the following—Ma Qian (Advisory Editor), Christina Gifford (Elsevier Acquisitions Editor), Heather Cain (Elsevier Editorial Project Manager) Jim Sears, Ryan DeHoff, Richard Grylls, Jessica Nehro, Anders Hultman, Scott Thompson, Laura Kinkopf, Michael Cloran, David Whittaker, and Karl D'Ambrosio.

CHAPTER 1

The Additive Manufacturing of Titanium Alloys

ABBREVIATIONS AND GLOSSARY

3D	three dimensional
AM	additive manufacturing
CAD	computer aided design
DED	directed energy deposition
DMD	direct metal deposition
DMLS	direct metal laser sintering
EBM	electron beam melting
GE	General Electric Corporation
LENS	laser engineered net shaping
PBF	powder bed fusion
P/M	powder metallurgy
SL	stereolithography

1.1 INTRODUCTION

1.1.1 Titanium Alloys and Their Importance

Titanium alloys are among the most important of the advanced materials that are key to improved performance in aerospace and terrestrial systems (Figs. 1.1–1.4) and is even finding applications in the cost conscience auto industry.[1–5] These applications result from the excellent combinations of specific mechanical properties (properties normalized by density) and outstanding corrosion behavior[6–11] exhibited by titanium alloys. However, negating its widespread use is the high cost of titanium alloys compared to competing materials (Table 1.1).

The high cost of titanium compared with the other metals shown in Table 1.1 has resulted in the yearly consumptions as shown in Table 1.2.

Additive Manufacturing of Titanium Alloys. DOI: http://dx.doi.org/10.1016/B978-0-12-804782-8.00001-X

Figure 1.1 The Boeing 787 Commercial airplane contains 20% titanium. Source: © The Boeing Company.

Figure 1.2 GE Aviation's GEnx is an advanced dual rotor, axial flow high bypass gas turbine engine for use on Boeing's 787 and 747-8 aircraft and features titanium allow compressor blades and disks. Source: GE Aviation.

Figure 1.3 Author Dr. F.H (Sam) Froes is shown in front of The Guggenheim Museum in Bilbao, Northern Spain, which is sheathed with titanium sheet.

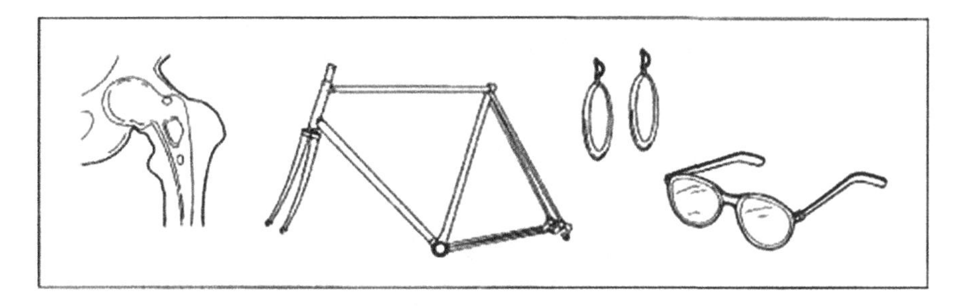

Figure 1.4 Titanium is used for a wide variety of items, such as bike frames, hip implants, eyeglass frames, and earrings.

Table 1.1 Cost of Titanium: A Comparison[a]

Item	Material ($/lb)		
	Steel	Aluminum	Titanium
Ore	0.02	0.01	0.22 (rutile)
Metal	0.10	1.10	5.44
Ingot	0.15	1.15	9.07
Sheet	0.30–0.60	1.00–5.00	15.00–50.00

[a]2015 Contract prices. The high cost of titanium compared to aluminum and steel is a result of (a) high extraction costs and (b) high processing costs. The latter relates to the relatively low processing temperatures used for titanium and the conditioning (surface regions contaminated at the processing temperatures, and surface cracks, both of which must be removed) required prior to further fabrication.

Table 1.2 Metal Consumption

Structural Metals	Consumption/Year (10^3 Metric Tons)
Ti	50
Steel	700,000
Stainless steel	13,000
Al	25,000

1.1.2 Challenges to Expanding the Scope of Titanium Alloys

In publications over the past few years[1–29] the cost of fabricating various titanium precursors and mill products has been discussed (very recently the price of TiO_2 has risen to US$2.00/lb and $TiCl_4$ to US $0.55/lb). The cost of extraction is a small fraction of the total cost of a component fabricated by the cast and wrought (ingot metallurgy) approach (Fig. 1.5). To reach a final component, the mill products shown in the figure must be machined, often with very high buy-to-fly

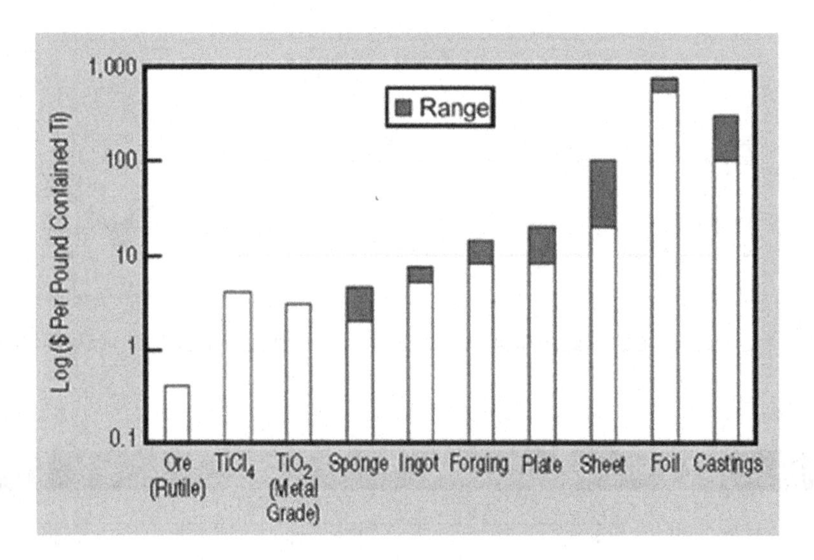

Figure 1.5 Cost of titanium at various stages of a component fabrication.

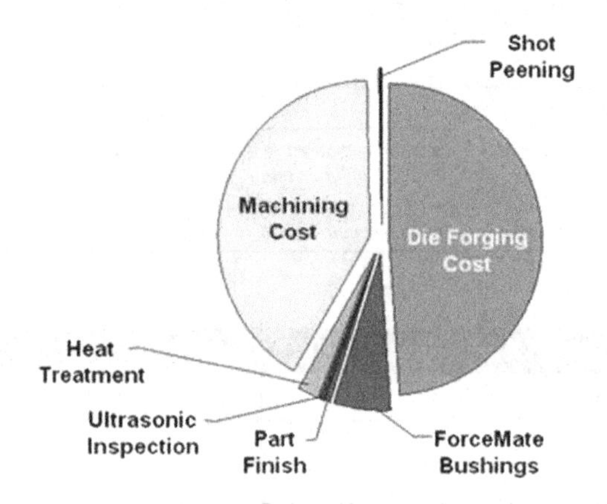

Figure 1.6 Boeing 787 side-of-body chord, manufacturing cost breakdown. Source: Courtesy of Boeing.

ratios (which can reach as high as 40:1). The generally accepted cost of machining a component is that it doubles the cost of the component (with the buy-to-fly ratio being another multiplier in cost per pound), as seen in Fig. 1.6. Fig. 1.7 illustrates how the machining of titanium has evolved, with rough machining showing a much greater improvement than the much more precise and more expensive final machining. Thus, while improvements in the machining of titanium have occurred, anything that can be done to produce a component which is closer to the final configuration will result in a cost reduction—hence the attraction of near-net shape components.

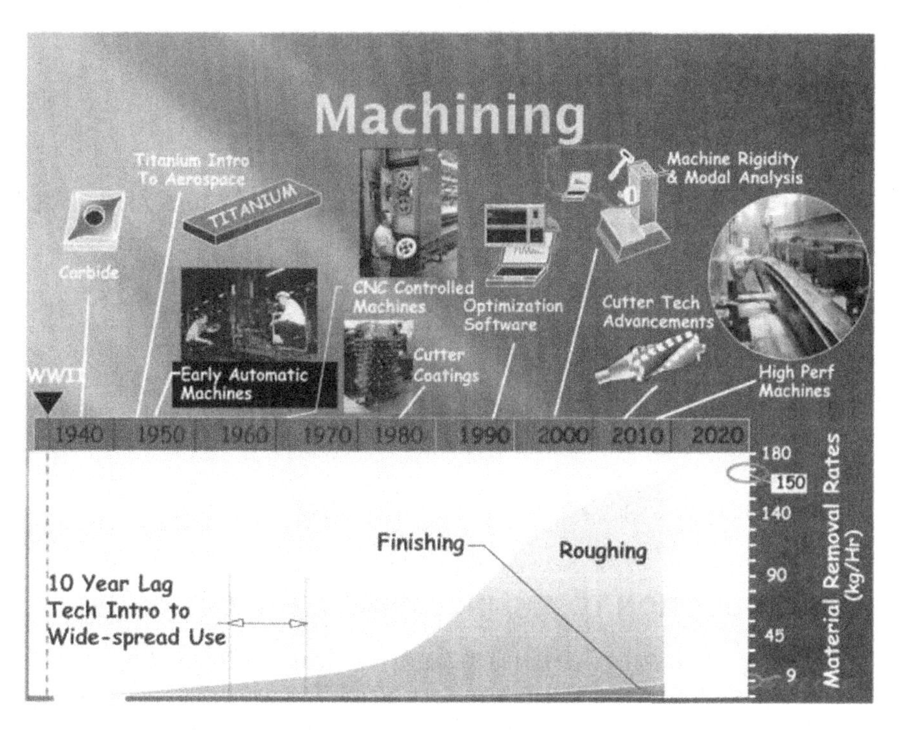

Figure 1.7 A historical perspective of the machining of titanium showing how various advancements have reduced the cost of machining. The light gray region represents the progress in rough machining and the brown(dark gray in print version) region shows how the final (much more precise and therefore costly) machining has evolved. Source: Courtesy of Boeing.

The high cost of conventional titanium components has led to numerous investigations of various potentially lower cost processes,[1–3] including powder metallurgy (P/M) near-net shape techniques.[1–29] In this book, one P/M near-net shape technique called additive manufacturing (AM) will be reviewed with the emphasis on the "work horse" titanium alloy Ti-6Al-4V, which accounts for more than 70% of the alloy titanium produced world-wide.[27] This technique (described below) is receiving a lot of attention from the US Navy who envision a future use of the approach aboard a carrier where parts can be rapidly fabricated for immediate use on the battle group that the carrier is supporting.[30] The various approaches to AM are presented, followed by some examples of components produced by AM. The microstructures and mechanical properties of Ti-6Al-4V produced by AM are listed and shown to compare very well with cast and wrought product. Finally the economic advantages to be gained using the AM technique compared to conventionally processed material are presented.

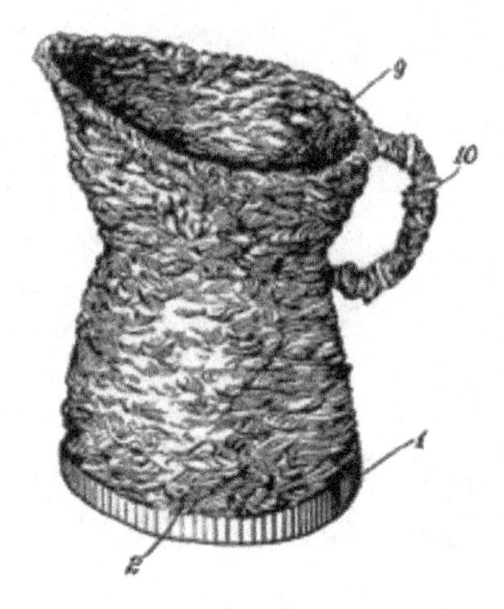

Figure 1.8 A typical decorative article as described by Baker's 1920 patent.[31]

1.2 INTRODUCTION TO AM, INCLUDING A HISTORY OF AM

The history of metal AM dates back to at least 1920 by Baker (US patent, 1,533,300) who used an electric arc and metal electrode to form walled structures and decorative articles (Fig. 1.8).[31] Today, directed energy deposition (DED) techniques, such as direct metal deposition (DMD), light engineered net shaping (LENS), or direct manufacturing (DM), are based on similar ideas but integrate layered manufacturing concept to create parts directly from computer-aided design (CAD) data. However the concept of layered manufacturing finds its roots from two different technologies that started in the 19th century: topography and photo sculpture.

As early as 1892, Blanther patented a technique for making a mold for topographical relief maps using impressions of topographical contour lines on a series of wax plates, cutting these wax plates on these lines, and stacking them to create raised relief map of paper. In 1972 Matsubara proposed a process using photopolymer resin coated onto graphite powder/sand, spread into a layer, selected areas of the layer heated, and hardened using a mercury vapor lamp and the remaining area dissolved to create sheets with defined geometry, which were then stacked together to form a casting mold. In 1974, utilizing a similar stacking technique, DiMatteo produced three-dimensional (3D) shapes from contour milled metallic sheets that were then joined in layered fashion by adhesion, bolts, or tapered rods (Fig. 1.9A). In 1968 Swainson proposed a process to

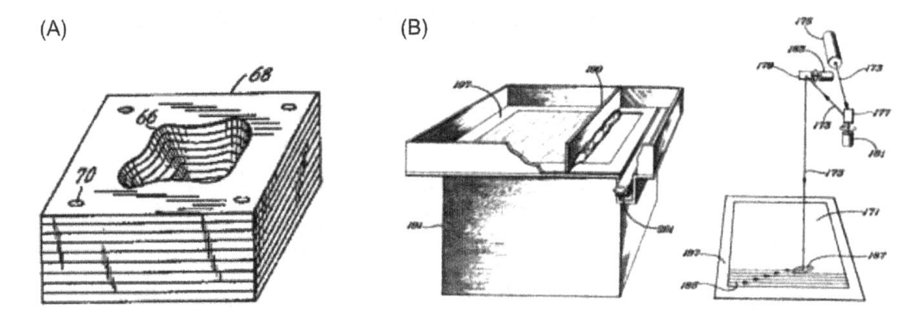

Figure 1.9 (A) Laminated mold from DiMatteo's 1974 patent[32] and (B) schematic drawing showing Householder's invention.[33]

directly fabricate a plastic pattern by selective 3D polymerization of a photosensitive polymer at the intersection of two laser beams. This was followed by work from Ciraud (1972), Householder (1979) (Fig. 1.9B), Kodama (1981), Herbert (1982), Hull (1984), and Deckerd (1986) paving the way for modern 3D printing technologies that are based on powder bed fusion (PBF) technologies.[34] Patents by Ciraud (1972), Arcella and Lessmann (1989), Jeantette et al. (1996), and Koch and Mazumder (1998) form the basis of modern DED technologies.[34] More details about the history of AM and 3D printing as well as commercialization of this industry can be found elsewhere.[34,35]

1.3 BRIEF INTRODUCTION OF VARIOUS AM TECHNOLOGIES

All AM technologies are based on the common principle of slicing a solid model into multiple layers, creating a tool path for each layer, uploading this data in the machine, and building the part up layer by layer following the sliced model data using a heat source (laser, electron beam, electric arc, or ultrasonic energy, etc.) and feed stock (metal powder, wire or thin metal sheet, etc.). ASTM F2792-12a categorizes all AM technologies into a broad group of seven categories: binder jetting, DED, material extrusion, material jetting, PBF, sheet lamination, and vat polymerization.[36] Of these seven categories, only four involve metal processing—DED, PBF, sheet lamination, and binder jetting—and only the first three of these four have been used for processing titanium and its alloys. While all three categories, DED, PBF, and ultrasonic consolidation have demonstrated capability of producing metal parts, their approach and capabilities vary significantly. The details of their working principles, capabilities, merits, and demerits will be discussed in Chapter 3, Additive Manufacturing Technology.

1.4 AM OF TITANIUM IN THE FOREFRONT OF THE METAL AM INDUSTRY

Commercial efforts of AM started with introduction of stereolithography (SL) and formation of the 3D Systems Corporation in 1987. This was followed by formation of EOS GmbH in 1990, Stratasys in 1991, DTM Corp. in 1992, Fockele & Schwarze (F&S) in 1994, Z Corp in 1996, and subsequently many other companies. While most of this effort was using polymeric materials, commercialization of metal AM started with DTM. DTM launched a metal sintering system, by laser sintering polymer-coated metal powders in the sinter station to form a green part, followed by a furnace process to remove the polymer, bond the metal matrix, and infiltrate it with a secondary metal to remove the porosity. In contrast to this, EOS developed a direct metal laser sintering (DMLS) process where metal powders were directly sintered using a moving laser beam.[37]

Titanium AM efforts began in 1997 at the Aeromet Corporation which focused on a laser-based DED technology for large aerospace components[38] and Arcam AB using its patented electron beam melting (EBM) technology to produce medical components. Arcam's continued work with Adler Ortho Group resulted in Conformité Européenne, meaning "European Conformity" (CE-certification) of EBM manufactured titanium hip implants in 2007 marking a significant step in titanium AM. In 1998 commercialization of Sandia National Laboratory-developed Laser Engineered Net Shaping (LENS) by Optomec Inc. and the University of Michigan-developed DMD by POM Group brought further thrust to metal AM and processing titanium. Due to economic reasons, early efforts on metal AM were focused on expensive parts and components, with aerospace and medical industry being a natural fit. This resulted in a major focus on titanium and its alloys, besides other expensive alloys. These efforts on titanium processing were soon followed by other companies such as EOS, Concept Lasers, MTT, SLM Solutions, Sciaky, Solidica, and more.

REFERENCES

1. Froes FH (Sam). Powder metallurgy of titanium alloys. In: Chang I, Zhao Y, editors. *Advances in powder metallurgy*. Philadelphia, PA: Woodhead Publishing; 2013. p. 202.

2. Froes FH (Sam), Imam MA, Fray D, editors. *Cost affordable titanium*. Warrendale, PA: TMS; 2004.

3. Gungor MN, Imam MA, Froes FH (Sam), editors. *Innovations in titanium technology*. Warrendale, PA: TMS; 2007.

4. Imam MA, Froes F H (Sam), Dring KF, editors. *Cost-affordable titanium III*. Switzerland: Trans Tech Publications Ltd; 2010.

5. Imam MA, Froes FH (Sam), Reddy RG, editors. *Cost affordable titanium IV*. Switzerland: Trans Tech Publications; 2013.

6. Froes FH (Sam). Titanium powder metallurgy: developments and opportunities in a sector poised for growth. *Powder Metall Rev* 2013;**2**(4):29–43 Winter.

7. Dutta B, Froes FH. Additive manufacturing of titanium alloys. *Adv Mater Process* February 2014;**172**(2):18–23.

8. Froes FH (Sam), editor. *Titanium physical metallurgy, processing and applications*. Materials Park, OH: ASM; February 2015.

9. Froes FH (Sam). Titanium alloys: alloy designation system [to be published] In: Hashmi S, editor. *Module in materials science and engineering*. Kidington, UK: Elsevier Publishing; 2015

10. Froes FH (Sam). "Titanium alloys: thermal treatment and thermomechanical processing [to be published] In: Hashmi S, editor. *Module in materials science and engineering*. Kidington, UK: Elsevier Publishing; 2015

11. Froes FH (Sam). Titanium alloys: properties and applications [to be published] In: Hashmi S, editor. *Module in materials science and engineering*. Kidington, UK: Elsevier Publishing; 2015

12. Froes FH (Sam). Titanium alloying [to be published] In: Hashmi S, editor. *Module in materials science and engineering*. Kidington, UK: Elsevier Publishing; 2015

13. Froes FH. Introduction: developments to date. In: Qian M, Froes FH (Sam), editors. *Titanium powder metallurgy*. Amsterdam: Elsevier and Science Direct; 2015. p. 1–19.

14. Yolton CF, Froes FH. Conventional titanium powder production. In: Qian M, Froes FH (Sam), editors. *Titanium powder metallurgy*. Amsterdam: Elsevier and Science Direct; 2015. p. 21–32.

15. Froes FH. Research based titanium powder processes. In: Qian M, Froes FH (Sam), editors. *Titanium powder metallurgy*. Amsterdam: Elsevier and Science Direct; 2015. p. 95–9.

16. Samarov V, Seliverstov D, Froes FH. Titanium components manufacture from prealloyed powder using hot isostatic pressing (HIP)". In: Qian M, Froes FH (Sam), editors. *Titanium powder metallurgy*. Amsterdam: Elsevier and Science Direct; 2015. p. 313–36.

17. Dutta B, Froes FH. Additive manufacturing of titanium. In: Qian M, Froes FH (Sam), editors. *Titanium powder metallurgy*. Amsterdam: Elsevier and Science Direct; 2015. p. 447–68.

18. Whittaker D, Froes FH. Current and future markets for titanium powder metallurgy. In: Qian M, Froes FH (Sam), editors. *Titanium powder metallurgy*. Amsterdam: Elsevier and Science Direct; 2015. p. 579–600.

19. Froes FH, Qian M. Titanium powder metallurgy for the future. In: Qian M, Froes FH (Sam), editors. *Titanium powder metallurgy*. Amsterdam: Elsevier and Science Direct; 2015. p. 601–8.

20. Froes FH (Sam), Dutta B. The additive manufacturing of titanium alloys. In: Proceedings of the light metals conference, Kwa Maritane, South Africa. Switzerland: Transtech Publishing; 2015.

21. Froes FH (Sam), Dutta B. The additive manufacturing of titanium alloys. To be published in the Proceedings of the World Conference on Titanium, San Diego, CA; 2015.

22. Froes FH, Eylon D, Bomberger H, editors. *Titanium technology: present status and future trends*. Dayton, OH: TDA; 1985.

23. Froes FH (Sam), Yau TL, Weidenger HG. Titanium, zirconium and hafnium. In: Matucha KH, editor. *Materials science and technology—structure and properties of nonferrous alloys*. Weinheim, FRG: VCH; 1996. p. 401 [chapter 8].

24. Froes FH (Sam). Titanium. In: Bridenbaugh P, editor. *Encyclopedia of materials science and engineering*. Oxford, UK: Elsevier; 2000 [chapters 3.3.5a–3.3.5e].

25. Froes FH (Sam). Titanium alloys. In: Weasel JK, editor. *Handbook of advanced materials*. New York, NY: McGraw-Hill Inc.; 2000 [chapter 8].

26. Froes FH (Sam). Titanium metal alloys. In: Ellis J, editor. *Handbook of chemical industry economics, inorganic*. New York, NY: John Wiley and Sons Inc.; 2000.

27. Boyer RR, Welsch G, Collings EW, editors. *Materials properties handbook: titanium alloys*. Materials Park, OH: ASM Int.; 1994.

28. Froes FH, Eylon D. Powder metallurgy of titanium alloys. *Int Mater Rev* 1990;**35**:162.

29. Froes FH, Suryanarayana C. Powder processing of titanium alloys. In: Bose A, German RM, Lawley A, editors. *Reviews in particulate materials*, vol. 1. Princeton, NJ: MPIF; 1993. p. 223.

30. Defence News, June 10, 2013. p. 24.

31. Baker R., US Patent 1,533,300; 1925.

32. DiMtteo P., US Patent 3,932,923; 1976.

33. Householder R.F., US Patent 4,247,508; 1981.

34. Beaman JJ, Barlow JW, Bourell DL, Crawford RH, Marcus HL, McAlea KP. *Solid freeform fabrication: a new direction in manufacturing*. New York: Springer; 1997. p. 1–21.

35. Terry Wohlers, Wohlers Report 2014.

36. ASTM F2792-12a.

37. Shellabear M, Nyrhilä O. In: PresentedPlease provide chapter title in Ref. [37]. at LANE 2004 conference, Erlangen, Germany; September 21–24, 2004.

38. Lutjering G, Williams JC. *Titanium*. Berlin: Springer; 2003. p. 95.

CHAPTER 2

Raw Materials for Additive Manufacturing of Titanium

ABBREVIATIONS AND GLOSSARY

ADMA	Advanced Materials (Corporation)
AM	additive manufacturing
ASTM	American Society for Testing Materials
ATI	Allegheny Technologies Incorporated
AWS	American Welding Society
CSIRO	Commonwealth Scientific and Industrial Research Organization
DED	direct energy deposition
DM	direct manufacturing
EB	electron beam
FFC	fray, farthing, and chen (powder production process inventors)
HDH	hydride–dehydride (powder)
ITP	International Titanium Powder (Corporation)
MER	Materials and Electrochemical Research (Corporation)
PBF	powder bed fusion
P/M	powder metallurgy
PREP	plasma rotating electrode process
SMD	shaped metal deposition
WAAM	wire arc additive manufacturing

2.1 INTRODUCTION

While most additive manufacturing technologies use powder as the raw material, there are a few exceptions, such as direct manufacturing (DM) and shaped metal deposition (SMD) or wire arc additive manufacturing (WAAM) that use titanium wire and ultrasonic consolidation which uses sheet metal and/or thin foils of titanium as feed stock. Since the majority of the AM technologies including all powder bed fusion (PBF) and most of the direct energy deposition (DED)

Additive Manufacturing of Titanium Alloys. DOI: http://dx.doi.org/10.1016/B978-0-12-804782-8.00002-1

technologies use metal powder, more emphasis will be given to powder production and its requirements for AM usage.

2.2 TITANIUM POWDER PREPARATION TECHNIQUES

Table 2.1 shows the characteristics of the leading different types of titanium powders that are either available or under development today.[1–6] The oxygen level of the *hydride–dehydride* (HDH) powder can be reduced by deoxidizing with calcium.[1] The spherical powders exhibit good flow characteristics while the more angular powders do not flow as well. The angular HDH can be converted to a spherical morphology using the Tekna process.[1] Generally the more free-flowing spherical powders are preferred for AM, however angular powders have also been successfully processed by this technique (see later in this chapter).

Table 2.1 The Characteristics of Different Types of Titanium Powders[1–6]

Type/Process	Elemental or Prealloyed	Advantages	Status/ Disadvantages
Hunter Process (pure sodium)	Elemental	Low cost, excellent for cold press/sinter sinter	Limited availability High chloride
HDH[a] Kroll Process (pure magnesium)	Elemental	Lower cost Good compactability low chloride	
HDH powder produced from alloys ingot, sheet, or scrap	Prealloyed	Readily available	High cost
Atomized	Prealloyed	High purity Available	High cost
REP/PREP[b]	Prealloyed	High purity	High cost compactable
ITP/Armstrong	Both	Compactable moderate cost Potential for low cost	Processible
FFC Cambridge	Both	TBD	Developmental
MER[c]	Both	TBD	Developmental
CSIRO TiRO[d]	Both	TBD	Developmental
ADMA Products hydride powder	Both	Lower cost and better compactability	Semicommercial

[a]*Hydride–dehydride.*
[b]*Rotating electrode powder/plasma rotating electrode powder.*
[c]*MER Corp., Tucson, AZ.*
[d]*CSIRO Melbourne, Australia.*

2.2.1 Irregularly Shaped Powder

The cost of spherical titanium powder is a major obstacle in commercialization of AM technologies as part of mainstream manufacturing. The stringent quality requirements, such as shape and flow specifications, lead to only few powder manufacturing processes amenable to it. Therefore, there is a significant interest in production of low cost, generally angular-shaped, titanium powder that can be used in AM either as is or after conversion to a spherical morphology (see Fig. 2.6).

Six nonmelt processes for the production of angular powder appear to have the greatest potential for scale-up and a lower cost product than the spherical materials,[1–6] along with the additional hydride powder process that has been developed by Advance Materials (ADMA) Products also of potential commercial interest.[1–6] These six processes are the HDH method, the FFC Cambridge approach, the MER technique, the Commonwealth Scientific and Industrial Research Organization (CSIRO) method, the Chinuka process, and the ITP (International Titanium Powder)/Armstrong technique. However, these powders are not yet available commercially and their relative cost and processing characteristics are yet to be established.

An example of the process for production of an angular powder is the FFC Cambridge approach in which titanium metal is produced at the cathode in an electrolyte (generally $CaCl_2$) by the removal of oxygen from the cathode, as seen in Figs. 2.1 and 2.2.[1–6] This technique allows the direct production of alloys such as Ti-6Al-4V at a cost which could be less than the conventional Kroll process.[1] It is possible to generate a spherical

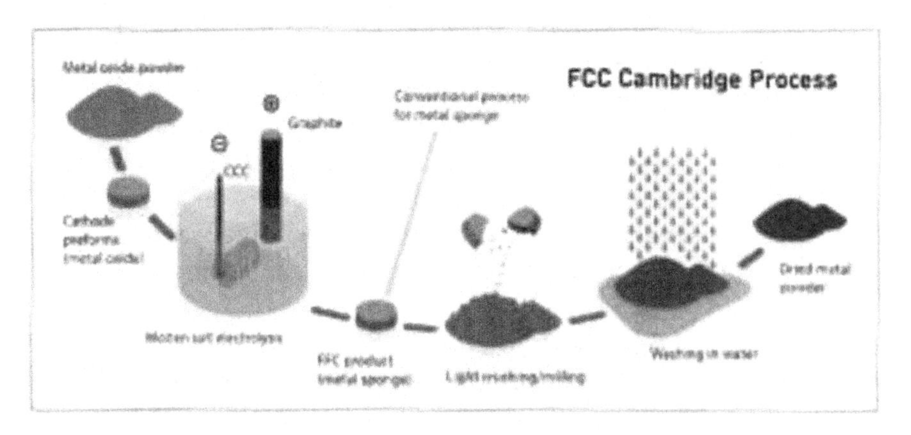

Figure 2.1 Schematic of the FFC Cambridge process. Source: Courtesy of Metalysis.

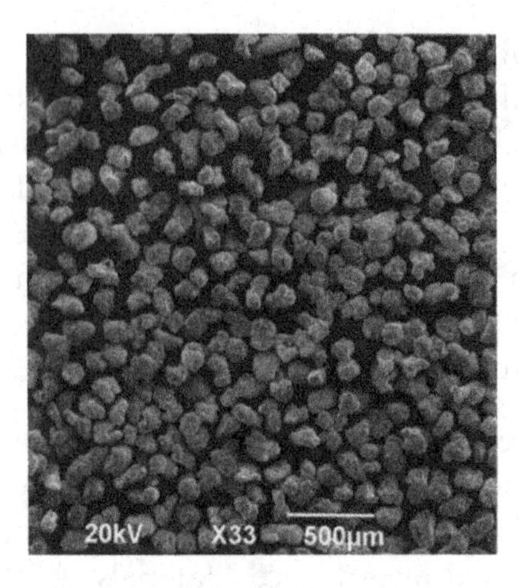

Figure 2.2 SEM Image of Ti-6Al-4V powder obtained by electrodeoxidation in Molten $CaCl_2$ at 950°C. Note the distorted spherical morphology of the Powder (the FFC Cambridge Process). Source: Courtesy of Ian Mellor, Metalysis.

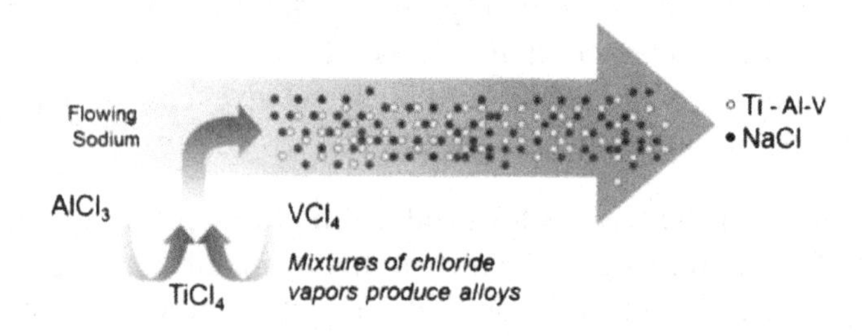

Figure 2.3 Schematic of the ITP/Armstrong Process. This process is claimed to be continuous, with a reaction efficiency of 100%, at a low temperature, low capital, low labor cost and can directly produce alloys such as Ti-6Al-4V. Source: Courtesy of Cristal Metals, Inc.

powder directly from the FFC process. The process allows routine production of 1000-μm diameter particles, and there is an active development program to lower this to the levels required for AM. The process is being commercialized by Metalysis in South Yorkshire, United Kingdom. The ITP/Armstrong process is shown schematically in Fig. 2.3 and an example of the powder produced in Fig. 2.4. Another example of an angular powder (HDH) is shown in Fig. 2.5. Examples of other processes to produce angular powders, and photographs of the products, are given in Ref. 6.

2.2.2 Prealloyed Spherical Powder

There are a number of processes which produce prealloyed spherical titanium powder (see Figs. 2.6 and 2.7 for examples),[1–6] with a recent increase in number as interest in titanium AM has significantly expanded.

Figure 2.4 Photomicrograph of angular Ti-6Al-4V powder produced by the ITP/Armstrong process. Source: Courtesy of Kerem Araci, Cristal Metals Inc.

Figure 2.5 Irregular hydride–dehydride Ti-6Al-4V powder. Source: Courtesy of Ametek Corp.

1. ATI Powder Metals (formerly Crucible Research Center). Spherical gas atomized alloy powder, 100 pounds capacity melting furnace, 50 pounds of −100/+325 (−150/+45 μm).
2. Advanced Specialty Metals. Spherical plasma rotating electrode process (PREP) −100/+325 (−150/+45 μm).
3. Raymor (now includes Pyrogenesis). Spherical plasma atomized Ti-6Al-4V powder. −450 to +60 mesh (−30/+250 μm) powder available, regular grade (0.20 O_2 max) Ti-6Al-4V, low oxygen (0.13 max). Arcam AB (a major fabricator of Titanium AM equipment) acquired the metal powder arm of Raymor and announced

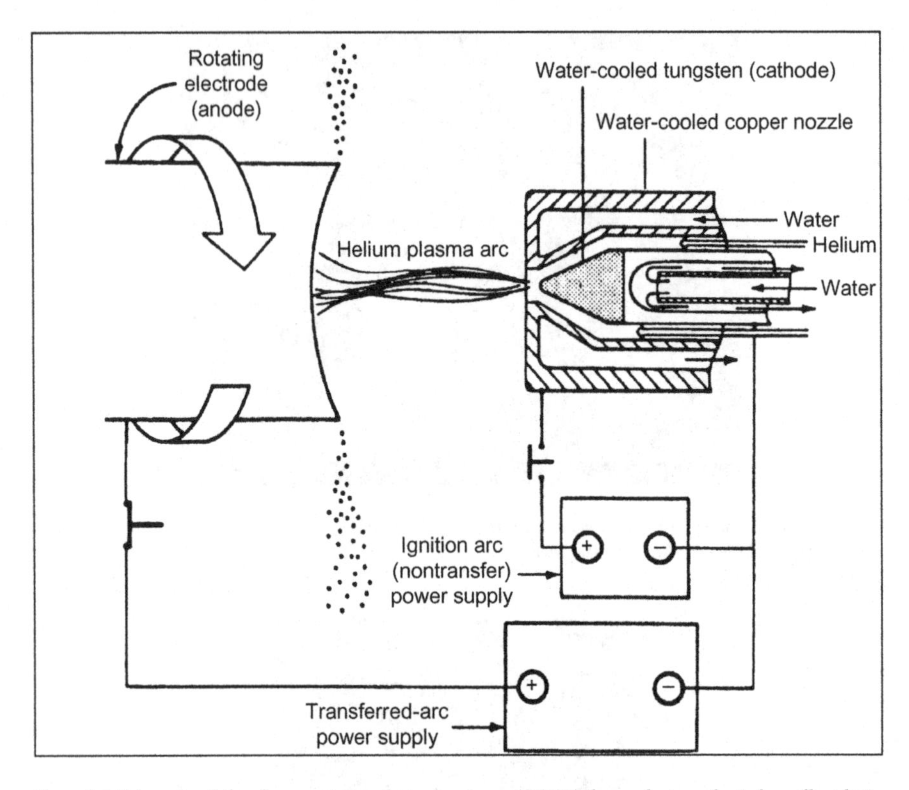

Figure 2.6 Schematic of the plasma rotating electrode process (PREP) for producing spherical prealloyed titanium powder.[1]

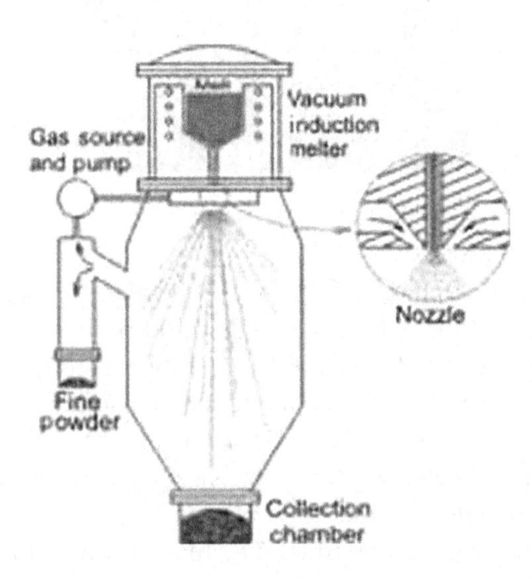

Figure 2.7 Schematic of the gas atomization (GA) process which produces spherical prealloyed titanium powder.

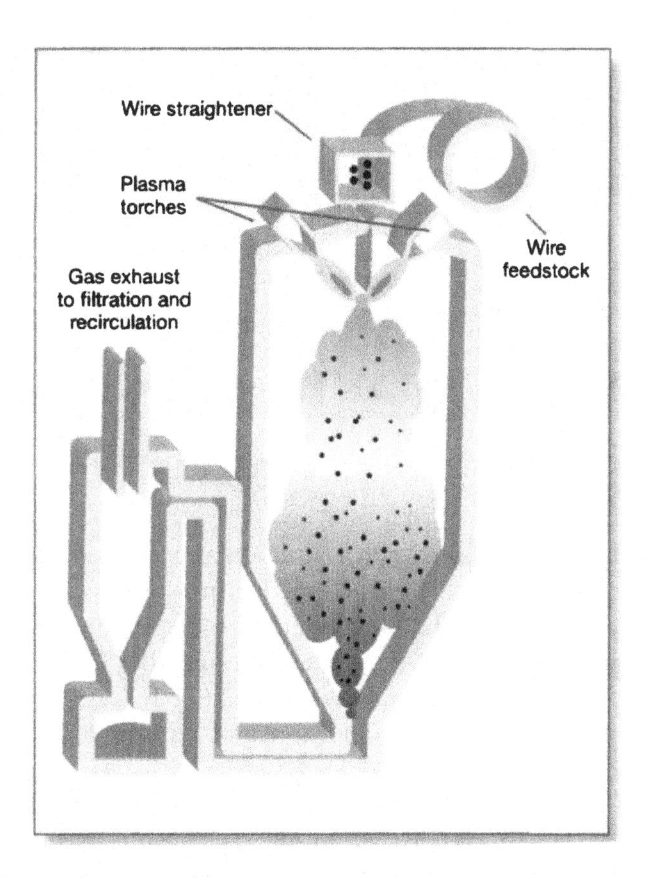

Figure 2.8 Schematic of the Arcam AB/Advanced Powders and Coating (AP&C) Plasma Atomization Process, using a wire feed.

that the metal powder subsidiary AP&C, based in Montreal, Canada, is building its fourth and fifth reactors using a wire feed (Fig. 2.8), adding significant capacity to its titanium powder manufacturing operations.

4. Baoji Orchid Titanium. Spherical PREP. Ti-6Al-4V −70/+325 (−210/+45 μm), 0.13 oxygen max.
5. ALD Vacuum Technologies. Spherical gas atomized Ti-6Al-4V electrode induction melting gas atomization.
6. Sumitomo Sitex. Gas atomized (Ti-6Al-4V, oxygen 0.08−0.13 wt%).
7. TLS Technik. Gas atomized Ti-6Al-4V with 0.13 oxygen. Ti-6Al-4V 100−270 mesh (53−150 μm).
8. Osaka Titanium offer Induction Melted Gas Atomized Titanium powder.

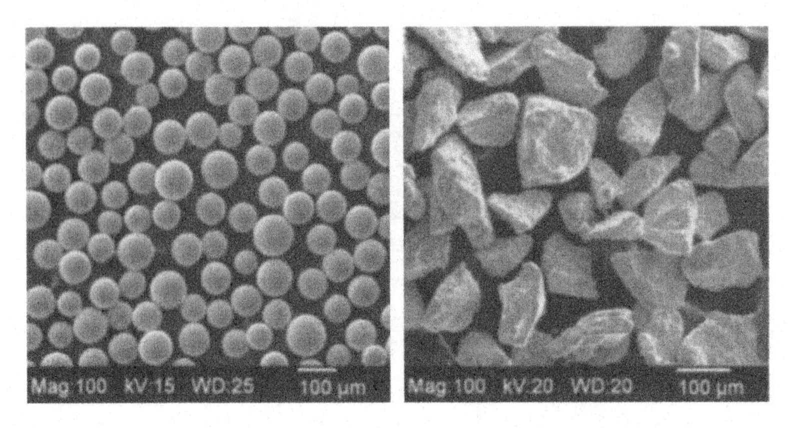

Figure 2.9 SEM photomicrograph of spherical powder (left) produced by processing angular HDH titanium powder (right) using the Tekna technique.

9. Tekna Induction plasma spherodization process converts irregular shaped titanium powders to a spherical morphology. Typically an irregular powder of −100/ + 400 mesh (−150/ +37 μm) is converted to a spherical powder of the same size range (but with a significant improvement in tap density and flow rate) (see Fig. 2.9).
10. LPW Technologies also offer the same services as Tekna using Tekna equipment.
11. Praxair Surface Technologies have announced (September 2015) that they plan to use close-coupled high pressure gas atomization to produce fine spherical titanium powder in large quantities, available by the end of the third quarter. They will be able to make a range of powder sizes: for laser PBF (10−45 μm) and electron beam PBF/powder fed technologies (45−150 μm). In addition to larger batches, the flexibility in particle size distribution is an advantage of the close-coupled high pressure gas atomization process.
12. Puris, located in West Virginia, United States, uses a Plasma Arc Melter to manufacture bars, or electrodes, for atomization. Producing electrodes in-house allows Puris to create custom chemistries with very precise elemental control, and facilitates achieving very low oxygen levels even for fine powders. Puris produces spherical gas atomized powder, 300,000 annual capacity, most common sizes are −140/ +325 μm and −325 μm mesh. Zero refractory in the system and the atomizer is constructed of Titanium to remove any risk of iron contamination (Puris is the only company that make those claims).

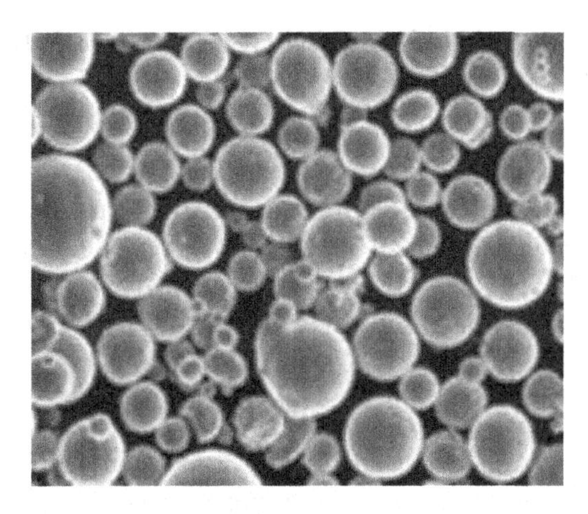

Figure 2.10 An example of spherical gas atomized titanium powder. Source: Courtesy of Osaka Titanium.

13. Hoeganeous. AncorTi is a gas-atomized spherical titanium powder for applications in additive manufacturing, metal injection molding, and hot isostatic pressing.
14. Carpenter Technology Corporation, Wyomissing, PA, United States, is reported to be spending an additional $23 million to add titanium furnace equipment to its new super alloy powder facility in Limestone County, Alabama. The purchase will raise Carpenter's investment in the plant to $61 million.

An example of a spherical powder is shown in Fig. 2.10. For further details on the prealloyed spherical titanium powders the reader is referred to Ref. 6.

2.3 WIRE FEED STOCK

Two DED processes that use metal wire as feedstock are SMD patented by Rolls Royce Corp and jointly developed with Cranfield University; and DM commercialized by Sciaky Inc. Both these processes use standard welding wires specified by AWS specifications and typical gage sizes of about 1.2 mm (0.05″) for SMD/WAAM[7] and 0.9–4.0 mm for DM.[8] According to AMS 4999A, the wire feedstock for titanium deposition process shall be wire conforming to AWS A5.16 ERTi-5 (Al content up to 7.5% is allowed, and a minimum of 1400 ppm oxygen).

2.4 ISSUES AND QUALIFICATION OF TITANIUM POWDERS FOR ADDITIVE MANUFACTURING

As noted earlier majority of the AM technologies use titanium or prealloyed titanium powders as feedstock. While titanium powder production has been well standardized over the years, its successful use in AM requires close control of the powder production process and handling of the powder. Below are certain key factors related to powder feedstock, which influence quality of AM parts. Details of characterization techniques that can be used for additive manufacturing technologies can be found elsewhere.[9,10]

1. *Size (and size distribution)*: In additive manufacturing, powder particle size determines the minimum part layer thickness, as well as the minimum buildable feature sizes on a part and surface finish in as-built condition. In addition, powder size distribution also plays a role in packing density for the PBF technologies. Typically PBF technologies use powder sizes ranging from 20 to 40 μm, while DED technologies use powder sizes ranging from 45 to 150 μm.
2. *Powder morphology*: Depending on the powder production process, powders can have various morphologies, including acicular, flake, granular, irregular, needle, nodular, platelet, plates, and spherical. Preferred powder morphology for AM are spherical shape and without any satellites. The morphology of powder particles determines the packing density of the powder and, hence, plays a major role in PBF technologies. The packing density will eventually determine the layer thickness and shrinkage in PBF processes. While powder morphology is less critical in DED, it is nevertheless important to maintain a consistent powder flow rate and therefore, a spherical or near spherical geometry is strongly recommended.
3. *Chemical composition*: Chemical composition of the powders play a major role as raw material chemistry determines final part chemistry and properties. The ASTM F42 Committee, who are responsible for developing AM Standards, have published chemistry specifications for most the commonly-used two Ti-alloys, namely, Ti-6Al-4V and Ti-6Al-4V-ELI (extra low interstitial) alloys for PBF processes.[11,12] It is to be noted that this Standard only requires chemistry compliance for the end part and allows for chemistry variation in the powder

form in order to adjust to the differences of various processes. It is worth noting that AM techniques involving electron beam melting operates in vacuum environment and therefore, leads to loss of elements, such as Al. Therefore the raw feedstock, such as Ti6-Al4-V powder for electron beam based processes often use additional amount of Al to compensate for the losses during processing. Raw material chemistry for Ti6-Al4-V alloy for DED processes can be found in the AMS 4999A specification.

4. *Flow*: Powder flow is a critical parameter for successful AM operations. Whether it is the recoater in PBF processes or powder delivery systems in DED processes, a consistent layer thickness depends on good and consistent powder flow. Powder morphology and surface characteristics, including surface contaminations, can affect the powder flow. For the purpose of obtaining consistent and repeatable powder flow, spherical powder shape without any satellites and surface contaminations are the preferred characteristics for powders in AM.
5. *Powder density*: Powder density is an important factor in AM. Hollow powders with or without gas entrapments can cause porosity effects in the end part and result in inferior part property. Fully dense powders are therefore highly recommended for AM.
6. *Powder handling and contamination*: As discussed earlier, powder contamination can cause impurities in the final part and effect properties. It is essential to exercise care while handling powder during transferring to powder hopper in machines or during sieving, etc.
7. *Effect of humidity*: Humidity plays a major role in AM. Moisture on the surface of the powder particles can cause porosity formation in the part. Therefore, it is essential to store the powder in dry places. If possible, powder drying prior to use is strongly recommended.
8. *Effect of powder recycling*: Powder recycling is a very important factor in AM processes. All PBF processes use only a fraction of the powder that is placed on the build plate and therefore powder recycling is mandatory for economic operation of these processes. Even though all titanium processing is carried out under inert environment or vacuum atmosphere, the powders are exposed to thermal effects from multiple layer builds and can have loss in elemental chemistry. ASTM F2924-14 and 3001-14 do not limit powder usage to any specific number of builds, and recommends to follow specific equipment guidelines for this purpose.

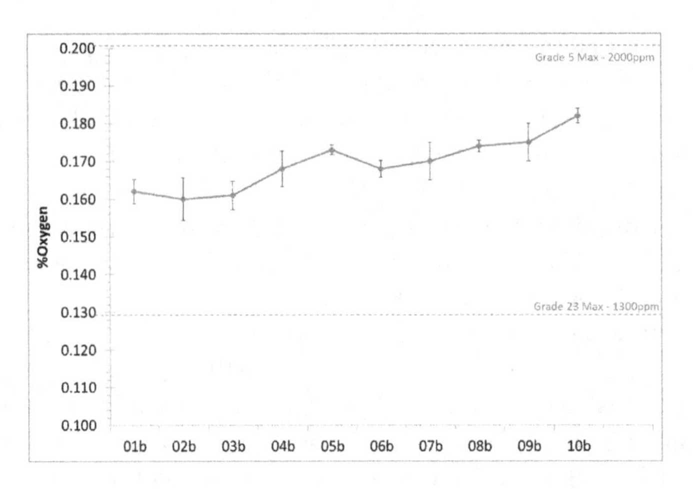

Figure 2.11 Effect of repeated runs through realizer powder bed fusion equipment on the oxygen content of Ti-6Al-4V.[13]

Tang et al. undertook an extensive study to evaluate the effect of powder reuse during electron beam melting (EBM) of Ti6-Al4-V alloys. Their results indicate that the powder morphology changes from smooth, spherical to rough, irregular after about 21 times of recycling. This change in powder morphology causes appreciable change in powder tap density (from 2.96 to 2.88 g/cc) and improves flowability (from 32.47 to 28.34 s/50 g). Enhanced powder flow during recycling is attributed to reduction. During this recycling process, powder chemistry was also affected, while V content changed marginally, Al content in the powder changed from 6.47% in virgin powder to 6.35% after 21 recycles and correspondingly, in the test samples from 6.14% to 5.93%. Loss of Al content is due to vaporization of Al in low partial pressure (2×10^{-2} Pa of He). Powder recycling is also associated with an increase in oxygen content from 0.08% in powder (0.07% in test sample) in virgin powder to 0.19% in powder (0.18% in test sample) after 21 recycles. Repeatable powder handling and exposure to air during recycling causes this additional oxygen pick up in the powder. As a result, the UTS increased from 920 to 1039 MPa, yield strength increased from 834 to 960 MPa and tensile elongation dropped from 16% to 15.5%. In other work using Realizer PBF equipment over 10 repeated runs, the oxygen content of Ti-6Al-4V was found to increase from 0.16 to 0.18 wt%, as seen in Fig. 2.11.[13]

REFERENCES

1. Froes FH, Eylon D. Powder metallurgy of titanium alloys. *Int Mater Rev* 1990;**35**:162.
2. Froes FH, Suryanarayana C. Powder processing of titanium alloys. In: Bose A, German RM, Lawley A, editors. *Reviews in particulate materials*, 1. Princeton, NJ: MPIF; 1993. p. 223.
3. Froes FH (Sam). Powder metallurgy of titanium alloys. In: Chang I, Zhao Y, editors. *Advances in powder metallurgy*. Philadelphia, PA: Woodhead Publishing; 2013. p. 202.
4. Froes FH (Sam). Titanium powder metallurgy: developments and opportunities in a sector poised for growth. *Powder Metall Rev* 2013;**Vol 2**(4):29–43 Winter.
5. Froes FH (Sam), editor. *Titanium physical metallurgy, processing and applications*. OH: ASM, Materials Park; February 2015.
6. Froes FH. Introduction: developments to date. In: Qian M, Froes FH (Sam), editors. *Titanium powder metallurgy*. Amsterdam: Elsevier and Science Direct; 2015.
7. Brandl E, Baufeld B, Leyens C, Gault R. Additive manufactured Ti-6Al-4V using welding wire: comparison of laser and arc beam deposition and evaluation with respect to aerospace material specifications. *Phys Proc* 2010;**5**:595–606.
8. <http://www.sciaky.com/additive-manufacturing/wire-am-vs-powder-am> [accessed 28.07.15].
9. Cooke A, Slotwinski J. Properties of metal powders for additive manufacturing: a review of the state of the art of metal powder property testing, NISTIR 7873. Available from: http://dx.doi.org/10.6028/NIST.IR.7873; July 2012.
10. Slotwinski JA, Garboczi EJ, Stutzman PE, Ferraris CF, Watson SS, Peltz MA. Characterization of metal powders used for additive manufacturing. *J Res Natl Inst Stand Technol* 2014;**Vol. 119**:460–93. Available from: http://dx.doi.org/10.6028/jres.119.018
11. ASTM standard F2924-14.
12. ASTM standard F3001-14.
13. Deffley R, LPW Technology Ltd, Private Communication; October 12, 2015.
14. Tang HP, Qian M, Liu N, Zhang XZ, Yang GY, Wang J. Effect of powder reuse times on additive manufacturing of Ti-6Al-4V by selective electron beam melting. *JOM* 2015;**67**(3):555–63. Available from: http://dx.doi.org/10.1007/s11837-015-1300-4

CHAPTER 3

Additive Manufacturing Technology

ABBREVIATIONS AND GLOSSARY

3D	three dimensional
ASTM	American Society for Testing Materials
AM	additive manufacturing
CAM	computer-aided manufacturing
CAD	computer-aided design
DED	directed energy deposition
DM	direct manufacturing
DMD	direct metal deposition
DMLS	direct metal laser sintering
DMDCAM	direct metal deposition computer-aided manufacturing
EBM	electron beam melting
ELI	extra low interstitial (composition)
GD&T	geometric dimensioning and tolerances
ID	internal dimension
LENS	laser energy net shaping
LM	laser melting
NIST	National Institute for Science and Technology
PBF	powder bed fusion
SLM	selective laser melting
SLS	selective laser sintering
STL	Standard Tessellation Language
UAM	ultrasonic additive manufacturing
WAAM	wire arc additive manufacturing

3.1 TECHNOLOGY OVERVIEW

Fig. 3.1 shows the process flow chart for a typical additive manufacturing (AM) process. It starts with selection of the part and determining part requirements. Once this has been accomplished, the design process begins with creating the CAD file and then the CAM toolpath for the AM equipment. At the other end of the process, given

Additive Manufacturing of Titanium Alloys. DOI: http://dx.doi.org/10.1016/B978-0-12-804782-8.00003-3

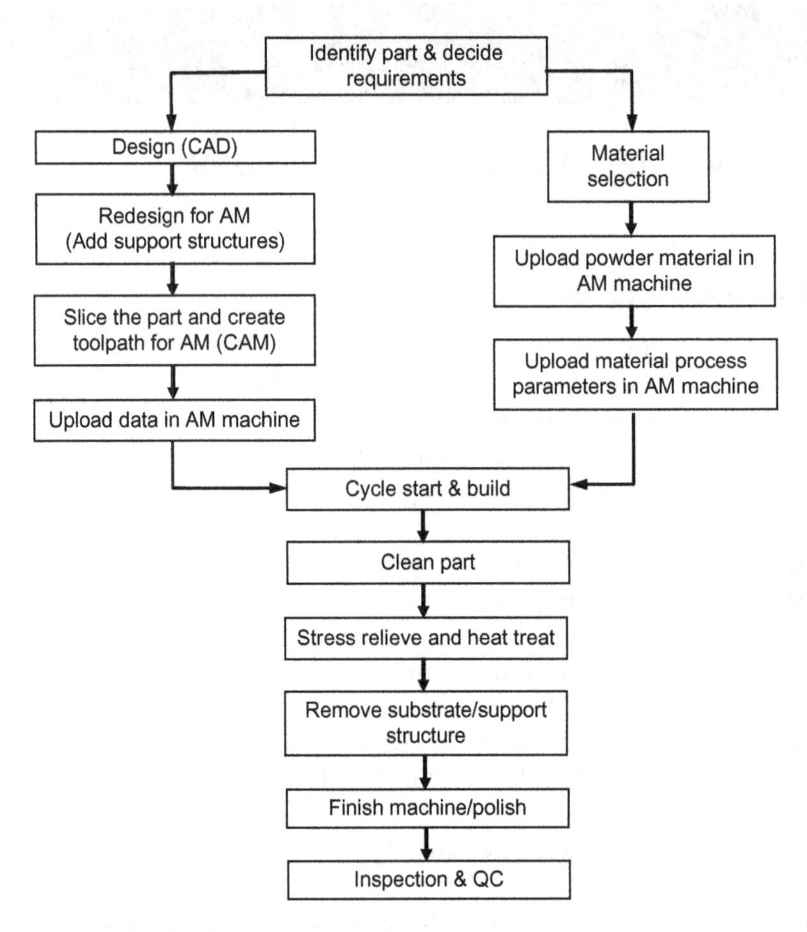

Figure 3.1 Process flow chart for a typical AM technology.

property requirements for the part suitable build material (in powder or wire form) is identified and appropriate AM process parameters are selected and data uploaded in the machine. The part is built up layer by layer following one layer at a time.[1–4] Once the process is over, the part is cleaned up, stress relieved, or heat treated following the part specification, substrate and/or support structures are removed, finish machined as per drawing specifications, inspected for compliance and its ready for use (Fig. 3.1). This section first addresses the design and creation of the computer toolpath and is followed by a discussion of various AM technologies for processing titanium with the main focus on two technologies: powder bed fusion (PBF) and directed energy deposition, followed by ultrasonic consolidation of sheet metal AM components.

3.2 SOFTWARE FOR AM

In principle, three-dimensional (3D) printing is based on taking a 3D geometry, slicing it into multiple layers, and creating a toolpath that will trace the part layer by layer, one layer at a time. 3D printing of metals has its roots in the stereolithographic process, invented by 3D Systems. Stereolithography was built on a surface file format, called STL (Standard Tessellation Language) and widely used in rapid prototyping and computer-aided manufacturing. Many of the metal-based 3D printing technologies use STL files as input. However, as STL represents the raw unstructured triangulated surface by the unit normal and vertices of the triangles using a 3D Cartesian coordinate system and does not contain any scale information, these files may not be suitable for complex operations and precision applications. Therefore many AM technologies are using solid models as input. Remanufacturing and/or surface coating using deposition based technologies (DED) pose additional challenges as they involve creating 3D layers as opposed to 2D layers that require five or six axis software for creation of the toolpath. Fig. 3.2 shows a typical deposition path simulated on a CAD model for a 5-axis deposition process using DMDCAM software.[5]

In PBF systems parts are fully built from scratch in a single setup. Therefore proper part geometry orientation is a critical step. In addition these systems also rely on support structures for building overhangs and designing the proper support structure is an essential part of a successful build strategy.[6] Various software, such as Magics from Materiallise, is

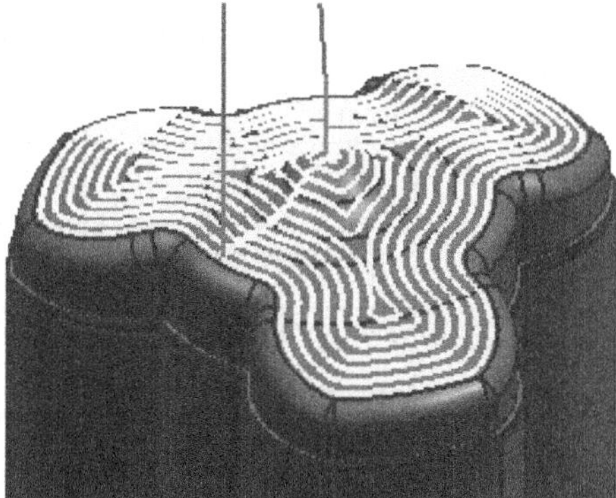

Figure 3.2 Left: CAD model of the part and process head. Right: Simulated tool path for 5-axis deposition using DMDCAM software. Source: Courtesy of DM3D Technology.

available that are dedicated to strategizing part orientation and building support structures for PBF systems.[7] Once the part is built, the support structures are machined off from the part.

3.3 PART BUILDING TECHNOLOGY

As mentioned in Chapter 1, The Additive Manufacturing (AM) of Titanium Alloys, ASTM classifies AM technologies into a broad group of seven categories: binder jetting, directed energy deposition (DED), material extrusion, material jetting, PBF, sheet lamination, and vat polymerization. Of these seven categories, only four involve metal processing: DED, PBF, sheet lamination, and binder jetting, and only the first three of these four have been used for processing titanium and its alloys, with the main focus being concentrated on PBF and DED technologies (Table 3.1). There are several

Table 3.1 Various AM Technologies for Processing of Titanium and Its Alloys[8–17]

AM Category	Technology	Company	Description
Directed energy deposition (DED)	Direct metal deposition (DMD)	DM3D Technology LLC (formerly POM Group)	Uses laser and metal powder for melting and depositing using a patented close loop process
	Laser engineered net shaping (LENS)	Optomec, Inc.	Uses laser and metal powder for melting and depositing
	Direct manufacturing (DM)	Sciaky, Inc.	Uses electron beam and metal wire for melting and depositing
	Shaped metal deposition or wire and arc additive manufacturing (WAAM)	Not commercialized yet (patented by Rolls Royce Plc.)	Uses electric arc and metal wire for melting and depositing
Powder bed fusion (PBF)	Selective laser sintering (SLS)	3D Systems Corp. (acquired Phenix Systems)	Uses laser and metal powder for sintering and bonding
	Direct metal laser sintering (DMLS)	EOS GmbH	Uses laser and metal powder for sintering, melting and bonding
	Laser melting (LM)	Renishaw Inc.	Uses laser and metal powder for melting and bonding
	Selective laser melting (SLM)	SLM Solutions GmbH	Uses laser and metal powder for melting and bonding
	LaserCUSING	Concept Laser GmbH	Uses laser and metal powder for melting and bonding
	Electron beam melting (EBM)	Arcam AB	Uses electron beam and metal powder for melting and bonding
Sheet lamination	Ultrasonic consolidation	Fabrisonic	Uses ultrasonic energy to consolidate layers of sheet metal and make parts

technologies under each category as branded by different manufacturers. While the PBF technologies enables building of complex features, hollow cooling passages, and high precision parts, these are limited by build envelop, single material per build, and horizontal layer building ability. In comparison, the DED technologies offer larger build envelop and a higher deposition rate, while their ability to build hollow cooling passages and finer geometry is limited. DMD and LENS technology also offer the ability to deposit multiple materials in a single build and the ability to add metal on existing parts. Commercially available AM technologies are based on three types of heat sources: laser, electron beam, and TIG arc, for the purpose of melting the feedstock (powder or wire). Laser-based systems operate under inert atmosphere (for titanium processing) in contrast to the vacuum environment of the electron beam systems. While the vacuum systems are more expensive, they offer the advantage of low residual stress as compared to laser based systems and electron beam processed parts can be used without any stress relieving operation. The effect of the heat source on the microstructure and mechanical properties is discussed in more detail in Chapter 4, Microstructure and Mechanical Properties.

3.3.1 Powder Bed Fusion

PBF technologies are based on the principles of laying down a layer of metal powder on the build platform and scanning the bed of powder with a heat source, such as a laser or an electron beam, that either partially or completely melts the powder in the path of the beam and resolidification and binding together of the powder as they cool down. Layer-by-layer tool path tracing is governed by the CAD data of the part being built. Fig. 3.3 shows a schematic diagram explaining the steps involved in this process.

- A substrate is fixed on the build platform.
- The build chamber is filled with inert gas (for laser processing) or evacuated (for electron beam processing) to reduce the oxygen level in the chamber to the desired level.
- A thin layer of the metal powder (20–200 μm thick depending on the technology and equipment) is laid down on the substrate and leveled to a predetermined thickness using a leveling mechanism.

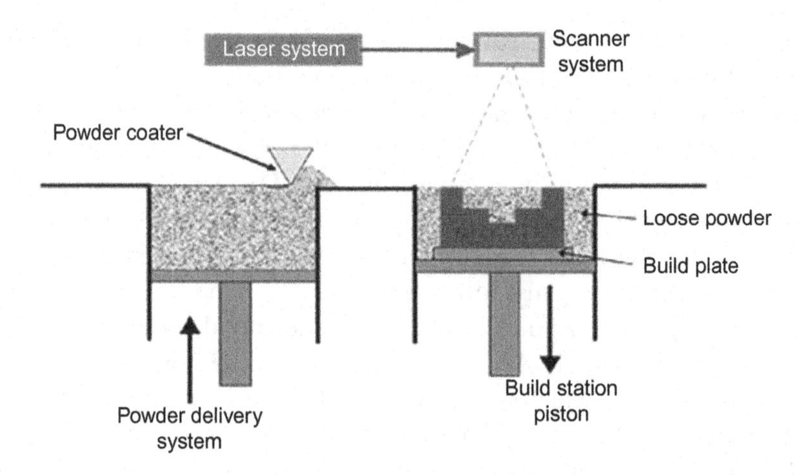

Figure 3.3 Schematic showing powder bed fusion technology. Source: Courtesy of Jim Sears.

- The laser or electron beam scans the powder bed surface following the toolpath pre-calculated from the CAD data of the component being built.
- The above process is repeated for the next and subsequent layers until the build is complete.

3.3.2 Directed Energy Deposition

DED technologies use material injection in to the meltpool instead of scanning on a powder bed (AMS specification 4999A for Ti-6Al-4V). Fig. 3.4 shows a schematic of the DMD technology (laser-based metal deposition). The process steps for the DED are:

- A substrate or existing part is placed on the work table.
- Similar to PBF, the machine chamber is closed and filled with inert gas (for laser processing) or evacuated (for electron beam processing) to reduce oxygen level in the chamber to the desired level (AMS 4999A specifies below 1200 ppm). The DMD process also offers local shielding and does not require inert gas chamber for less reactive metals than titanium, such as, steels, Ni-alloys, Co-alloys, etc.
- At the start of the cycle, the process nozzle with a concentric laser or electron beam is focused on the part surface to create a meltpool. Material delivery is in the form of powder through a coaxial nozzle (for laser) or through a metal wire with a side delivery (for electron beam). The nozzle moves at a constant speed and follows a

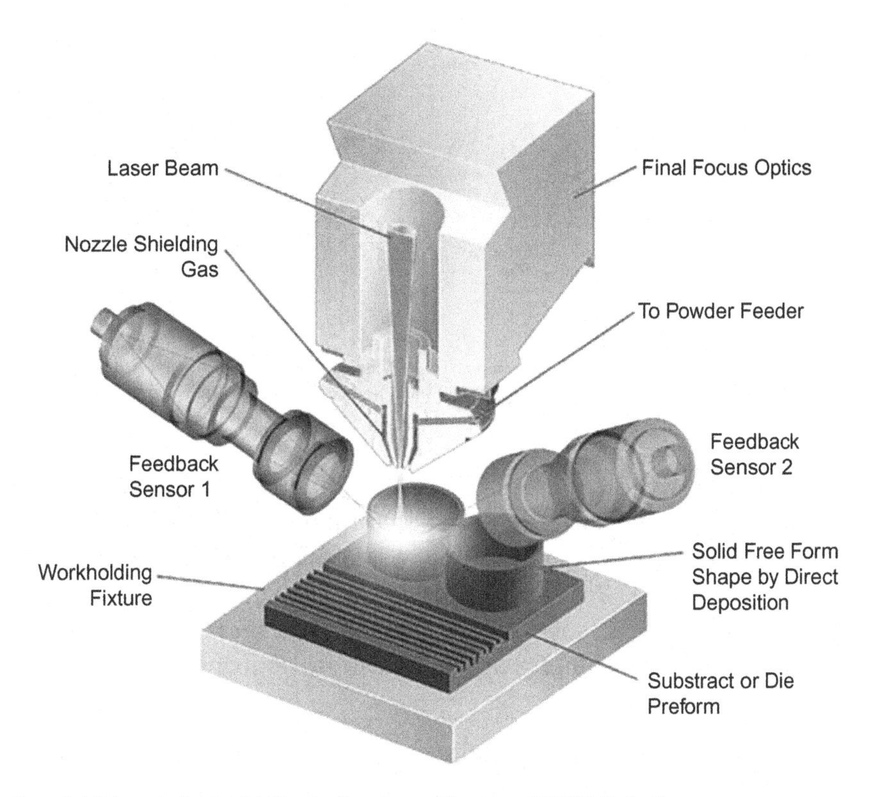

Figure 3.4 Schematic showing DMD technology. Source: Courtesy of DM3D Technology.

predetermined toolpath created from the CAD data. As the nozzle (tooltip) moves away the meltpool solidifies forming a layer of metal.

- Successive layers follow the same principle and build up the part layer by layer until completion.

3.3.3 Other AM Processes

Among other AM processes in the sheet lamination category, the ultrasonic additive manufacturing (UAM) process has been applied to process components containing titanium and aluminum. The UAM process involves building up solid metal objects through ultrasonically welding of a succession of metal tapes into a 3D shape, with periodic machining operations to create the detailed features of the resultant object. Fig. 3.5A shows a rolling ultrasonic welding system, consisting of an ultrasonic transducer, a booster, a (welding) horn, and a "dummy" booster. The vibrations of the transducer are transmitted, through the booster section, to the disk-shaped welding

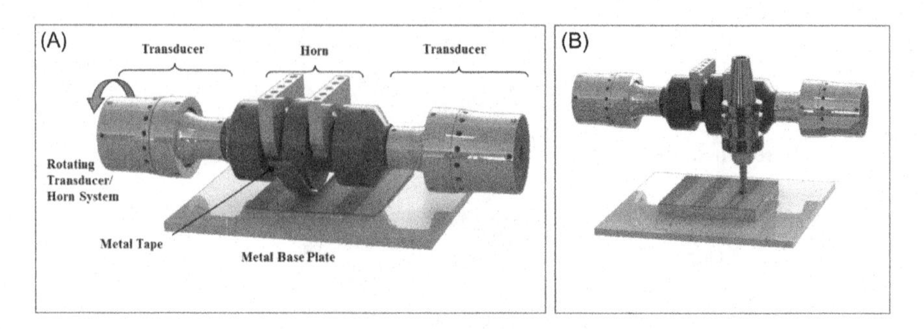

Figure 3.5 Ultrasonic additive manufacturing process. (A) Ultrasonic welding of aluminum and titanium tape; (B) Periodic machining operations. Source: Courtesy of Mark Norfolk, Fabrisonic LLC.

horn, which in turn creates an ultrasonic solid-state weld between the thin metal tape and base plate. The continuous rolling of the horn over the plate welds the entire tape to the plate. This is the essential building block of UAM. It is to be noted that the "horn" shown in Fig. 3.5A is a single, solid piece of metal that must be acoustically designed, so that it resonates at the ultrasonic frequency of the system (typically at 20 kHz).

Through welding a succession of tapes, first side-by-side to create a layer, and then one on top of the other (but staggered in the manner of bricks in a wall so that seams do not overlap), a 3D component is fabricated. During the build, periodic machining operations add features to the part, for example the slot in Fig. 3.5B, remove excess tape material and true up the top surface for the next stage of welds. Thus in this case, the so-called "AM" involves both additive and subtractive steps in arriving at a final part shape.

3.4 PROCESS CONTROL AND IN SITU MONITORING

All the AM processes involve a large number of variables including power of the heat source, speed, powder/wire feed rate (powder layer thickness in case of PBF processes), overlap of adjacent paths, gas flow rates, etc. A variation or fluctuation of any or a combination of these variables can affect the build process and result in poor part quality. So, it is essential to be able to monitor and possibly control as many variables possible on-the-fly in order to detect and avoid defects in the part. In recent years, there has been a lot of emphasis on process control and in situ process monitoring during the buildup process. In general, all the process monitoring and control variables can be placed into four major categories:

1. *Machine state monitoring*: Various OEMs for PBF technology are developing process monitoring and control (such as QMatmosphere from Concept Laser or EOSTATE from EOS) through monitoring the machine state, powder layer, and the melt pool. Machine state monitoring involves parameters such as the O_2 content in the chamber, gas filters, temperature inside the build chamber.
2. *Powder feed or powder layer thickness monitoring*: Since all AM processes are based on the concept of layer-wise manufacturing, precise control of layer thickness is an essential part of process control. Powder flow rate monitoring in DED systems is somewhat trivial and indirectly performed through monitoring and control of powder delivery system regulators. For example, the DMD process relies on powder delivery through a rotating powder feed shaft and feedback control electric motor regulates the powder feed rate to the meltpool. However, any variation from the powder delivery system to the meltpool will not be recorded in this process. In contrast, PBF systems have incorporated inspection systems of the powder layers following the powder recoating for each layer. This typically involves acquiring visible-light images of the entire bed, in which nonuniformities show up as contrasting features. If any anomaly is detected, the powder recoating process is repeated to ensure uniform powder distribution throughout the powder bed. Fig. 3.6 shows pictures taken from Concept Laser's QMcoating module depend on capturing visual images of the powder layer after

Figure 3.6 Concept Laser's QMcoating application. Red or darker color (dark gray in print version) shows excess powder. Left is without QMcoating module and right is with QMcoating module (Dunsky[18]). Source: Courtesy of Industrial Laser Solutions / Pennwell Corporation.

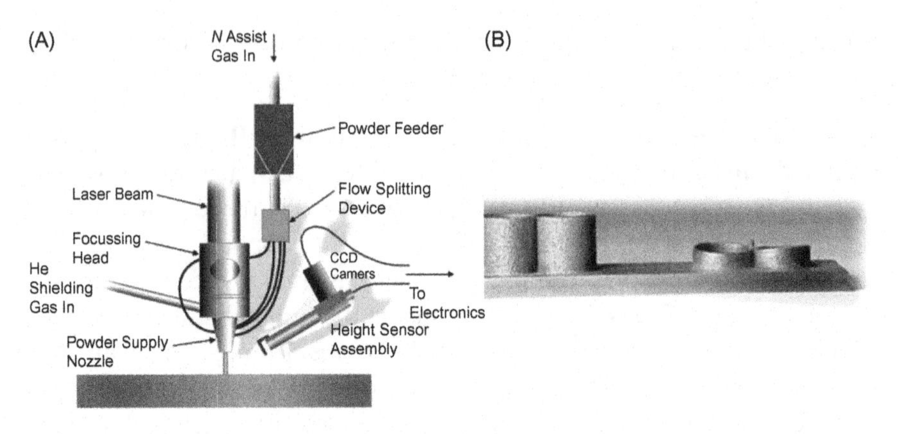

Figure 3.7 DM3D's patented closed loop feedback control system. (a) Schematic illustration. (b) Effect of process on a part with control (left) and without a control (right).[20]

the recoating process and detecting any anomaly through visual differences of contrast in the images.[18]

3. *Melt pool monitoring*: Direct sensing of the melt pool through photodiode or pyrometer or camera-based monitoring provides direct information about heat source and powder metal interaction. DM3D Technology employs a patented closed loop feedback control technology that uses a high-speed CCD cameras to measure the layer height during the deposition process and modulates laser power and/or nozzle speed in order to obtain the desirable layer thickness.[19] Fig. 3.7 shows a picture to illustrate the effect of the closed loop feedback control system. Two intersecting cylinders built with and without closed loop system demonstrates the elimination of excess build up height at the intersection with the help of the feedback system.

   DMLS technology employs a melt pool monitoring based on photo diode. Signals from the meltpool can be collected with off-axis sensors or on-axis sensors and the presence of a defect is reflected through an anomaly in the signal when compared to a standard reference signal (Fig. 3.8).[21]

   The Above concept of process monitoring can be taken one step further to process control. Kruth and Mercelis proposed a process control where melt pool dimensions can be calibrated for a known standardized sample to obtain the effect of various process parameters, such as laser power.[22] Once a relationship is established between the meltpool and input variables, this can be used for online compensation of any undesirable process variation.

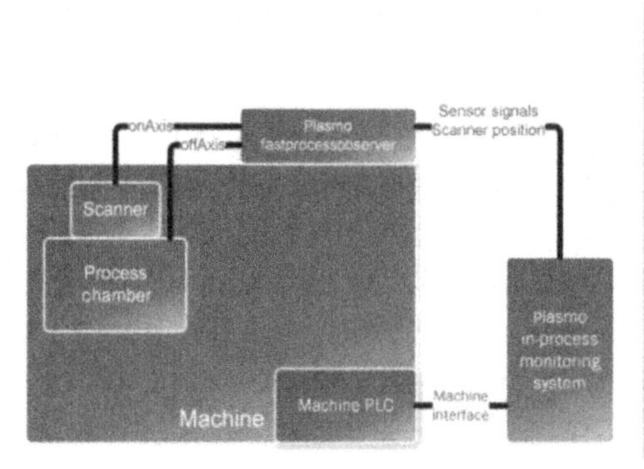

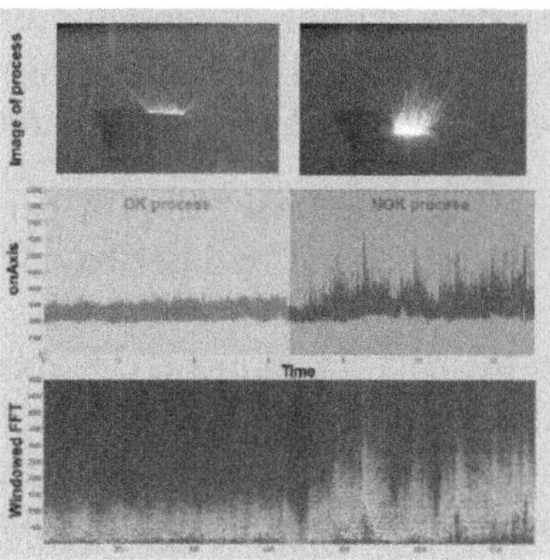

Figure 3.8 EOS's photodiode-based melt pool monitoring has been developed using Plasmo fast process observer.[21]

4. *Porosity monitoring*: Since the AM employs a layer-by-layer buildup strategy it also offers the opportunity of inspection after each layer is built. Arcam AB employs a process monitoring technology that is of great importance for highly demanding AM applications, for which the entire inspection workflow typically needs to be rethought.[23] LayerQam is a camera-based quality verification system for in situ process monitoring, integrated in the latest generation of Arcam EBM machines. The main purpose of LayerQam is to monitor porosity created during the melt process as the product is built layer by layer. LayerQam takes pictures of every layer in the EBM process in visible light with some extension into the InfraRed region. An image processing algorithm is implemented to analyze each melted layer and automatically generate a report postbuild with customized criteria. By stacking the pictures on top of each other, a 3D image similar to a CT scan can also be generated.

3.5 POSTPROCESSING OF AM PARTS

The first step after a part is built involves cleaning the part of residual powders through a shot peening or sand blasting process. Usually PBF parts processed using an electron beam require more time for cleaning due to stronger adherence of loose powder to the part. Once the part is cleaned, the next step is to separate the part from the substrate and removal of support structures, if they were used. Depending upon geometry and other requirements, this can be accomplished by simple bandsaw cutting, milling in a CNC machine and/or wire EDM cutting. A typical AM process is usually followed by heat treatment and secondary machining as required by the final part specification.

Heat treatment: All AM parts are amenable to any heat treatment. Parts processed by laser-based AM are usually followed with a stress relief treatment as rapid cooling laser processes induce residual stresses in the part. In contrast, electron beam–based AM parts do not require any stress relief. Depending on the final requirements, AM parts are often treated by h*ot isostatic pressing (HIP)*. Besides reducing the risk of undesirable porosities, a HIP process helps to break down the typical columnar microstructures that are associated with AM processes and allows for more equiaxed microstructure resulting in a more homogeneous microstructure and isotropic properties. This can be followed by any aging heat treatment as required by final part specification. ASTM F2924-14 and ASTM F 3001-14 specify heat-treatment for Ti-6Al-4V parts made using PBF systems, while AMS 4999A provides thermal treatment for Ti-6Al-4V parts using DED (laser based) processes.[24–26]

Postmachining: The surface finish and accuracy of AM parts depend on the process and technology. PBF processes use finer powder size as a raw material and smaller laser/electron beam spot size for processing, resulting in much better surface finish and higher accuracy of parts than DED processes. While almost all DED processed parts require some form of machining, PBF processed parts may or may not require finish machining depending on the final part requirements.[6] Sometimes, only sealing areas or mating surfaces are machined, leaving the rest of the surfaces untouched. Finishing processes such as slurry-based honing is widely used to improve surface finish of internal passages. Regular machining equipment, such as milling, can be used to finish machine AM parts; however, higher surface roughness of DED parts can pose challenges due to interrupted cuts on external surfaces. Grinding and/or EDM may be an alternative solution.

3.6 INSPECTION AND QUALIFICATION OF AM PARTS

Inspection and quality control of AM parts is an area of special interest and is currently under development. This becomes particularly critical as many of these parts are targeted toward aerospace and medical industries that demand higher qualification standards and material traceability. Wide acceptance of AM processes in industry depends very much on developing an affordable and rapid qualification and certification process for these parts. Predictive models, standardized process, materials and testing, and development of performance metrics and measurement methods are critical in order to develop this capability.

Current efforts to develop a qualification and certification process for AM parts follow modification and adaption of part qualification protocols that exist for other technologies. The ability to certify and qualify parts to existing specifications (eg, aerospace grade) in the field after secondary processing is critically important in order to accelerate the entry of AM as a mainstream manufacturing technology.[27] These efforts are aimed at replacing parts made using other manufacturing technologies with AM processes. This requires developing Standards for the AM processes, materials, and testing. The ASTM F42 Committee has been created with the goal of producing Standards for all AM processes. This Standard addresses material specifications for AM, feedstock handling, storing and disposal, machine monitoring and control, as well as testing of the parts produced using AM. Powder Standards include best practices for powder handling and storage to minimize contamination, reduce moisture absorption, and tracking of powder recycling, while the Standards also identify critical powder parameters that affect part quality and addresses methods to characterize these powder parameters. A part of the qualification process involves recording of all machine parameters during the process. Most of the commercial systems from companies, such as EOS, SLM Solutions, ARCAM, and Concept Laser allow process data collection modules. This data includes machine state parameters, run time, part details, etc., for the purpose of traceability. While the digital nature of these processes allows easy access to data collection, data storage and analysis of this very large amount of data poses a difficult challenge. This is currently being addressed by OEMs as well as various research institutes. NIST has published a detailed report listing all the testing methods that are applicable for AM parts.[28] As expected, with very few exceptions all standard material testing procedures are applicable for AM (Table 3.2).

Table 3.2 A List of Available Standards Related to PBF and DED Systems and Processes, Powders, and Testing of Parts

Designation	Title
Design	
ISO/ ASTM52915-13	Standard Specification for Additive Manufacturing File Format (AMF) Version 1.1
Materials and Processes	
F2924–14	Standard Specification for Additive Manufacturing Titanium-6 Aluminum-4 Vanadium with Powder Bed Fusion
F3001–14	Standard Specification for Additive Manufacturing Titanium-6 Aluminum-4 Vanadium ELI (Extra Low Interstitial) with Powder Bed Fusion
AMS 4999A	Standard Specification for Additive Manufacturing Titanium-6 Aluminum-4 Vanadium with Direct Metal Deposition (DED)
F3049–14	Standard Guide for Characterizing Properties of Metal Powders Used for Additive Manufacturing Processes
AMS 4998	Standard Guide for Characterizing Properties of Titanium-6 Alumnium-4 Vanadium Metal Powders Used in DED
Terminology	
F2792-12a	Standard Terminology for Additive Manufacturing Technologies
ISO/ ASTM52921-13	Standard Terminology for Additive Manufacturing-Coordinate Systems and Test Methodologies
Test Methods	
F2971–13	Standard Practice for Reporting Data for Test Specimens Prepared by Additive Manufacturing
F3122–14	Standard Guide for Evaluating Mechanical Properties of Metal Materials Made via Additive Manufacturing Processes
ISO/ ASTM52921–13	Standard Terminology for Additive Manufacturing-Coordinate Systems and Test Methodologies

ASTM standard for DED process is currently under development.

Future efforts in this area need to focus on new qualifications and certification Standards specially designed to exploit the nature of AM. As AM is a layer-by-layer buildup process, it offers a unique opportunity of online monitoring of the part quality.[29,30] The online part inspection can involve nondestructive (NDE) techniques based on X-ray or ultrasound or high definition digital images, as well as part distortion measurements, and residual stress measurements. Methods must be flexible and adaptive (rather than prescriptive) to allow for the broadest possible use. Multimaterial manufacturing ability is one of the major capabilities of AM technologies and future inspection techniques as well as qualification procedures need to address this aspect of AM.

While regular inspection techniques, destructive and NDE are very much applicable to AM parts, care must be taken to compensate for surface roughness of as-built AM parts when practicing these techniques. If AM parts are machined to surface specifications of the ASTM testing, standard testing techniques are adequate.[29] However, some AM applications are limited to machining access (such as cooling passages, narrow cavities or lattice structures, etc.) to sections of the part or the whole part. Sometimes, AM built parts are not machined to reduce cost of manufacturing. In order to simulate the performance of these parts, surface roughness must be taken into account while conducting any mechanical testing. Higher surface roughness of these parts as compared to standard machined surfaces may also pose challenges to some NDE techniques, such as ultrasonic inspection. This issue is particularly evident in DED built parts where the surface finish is coarser than PBF built parts. NDE techniques, such as density measurement, X-ray, and CT scans, are regularly performed for defect detection or porosity determination of AM parts.

Dimensional accuracy of AM parts is a topic of particular interest as many parts created by AM are not intended for any finish machining on many surfaces, for example scaffolding. Significant work is underway to understand distortion and dimensional accuracy of fine and intricate AM parts, including precision features such as small holes and thin walls.[31] NIST is actively working with ASTM F42 and the ISO's "Technical Committee 261 on Additive Manufacturing" to create test artifacts representing each different AM technology category in order to create an acceptance criteria for geometric dimensioning and tolerancing of each AM technology.[29] Once established, these Standards are expected to assist part certification and qualification process.

REFERENCES

1. Dutta B, Froes FH (Sam). Additive manufacturing of titanium alloys. *Adv Mater Proc* February 2014;18–23.

2. Froes FH, Dutta B. The additive manufacturing (AM) of titanium alloys. *Adv Mater Res* 2014;**1019**:19–25.

3. Dutta B, Froes FH (Sam). [chapter 24]: The additive manufacturing of titanium alloys. In: Qian M, Froes FH (Sam), editors. *Titanium powder metallurgy*. Elsevier Inc.; February 2015 Available from: http://dx.doi.org/10.1016/B978-0-12-800054-0.00024-1.

4. Froes (Sam) F.H, Dutta B. The additive manufacturing of titanium alloys. To be published in the Proceedings of the world conference on titanium, San Diego, CA; 2015.

5. DM3D Technology, private communication.

6. Moylan S, Slotwinski J, Cooke A, Jurrens K, Donmez MA. Lessons learned in establishing the nist metal additive manufacturing laboratory. Intelligent Systems Division, Engineering Laboratory, NIST, NIST Technical Note 1801. Available from: http://dx.doi.org/10.6028/NIST.TN.1801; June 2013.

7. <http://software.materialise.com/magics>.

8. <http://www.dm3dtech.com/index.php?option=com_content&view=article&id=86&Itemid=552> [accessed 17.11.15].

9. <http://www.optomec.com/3d-printed-metals/lens-printers/> [accessed 17.11.15].

10. <http://www.sciaky.com/additive-manufacturing/electron-beam-additive-manufacturing-technology> [accessed 18.11.15].

11. <http://www.phenix-systems.com/> [accessed 18.11.15].

12. <http://www.eos.info/additive_manufacturing/for_technology_interested> [accessed 18.11.15].

13. <http://www.renishaw.com/en/metal-additive-manufacturing-3d-printing--15240> [accessed 18.11.15].

14. <http://slm-solutions.us/> [accessed 18.11.15].

15. <http://www.concept-laser.de/en/technology.html> [accessed 18.11.15].

16. <http://www.arcam.com/technology/additive-manufacturing/> [accessed 18.11.15].

17. <http://fabrisonic.com/technology/> [accessed 18.11.15].

18. Dunsky C. Process monitoring in laser additive manufacturing. *Industrial Laser Solutions* September/October 2014;14–18

19. Koch J, Mazumder J. US Patent 6,122,564; September 2000.

20. Dutta B, private communication. DM3D Technology; July 2013.

21. Grünberger T, Domröse R. Optical in-process monitoring of direct metal laser sintering (DMLS). *Laser Tech J* 2014;**11**:40–2.

22. Kruth J-P, Mercelis P. Procedure and apparatus for in-situ monitoring and feedback control of selective laser powder processing. US Patent Application: US 2009/0206065 A1; August 20, 2009.

23. Bradshaw B, private communication, ARCAM AB; January 2016.

24. ASTM F2914-14, American Soc. For Testing of Metals; January 2012.

25. ASTM F3001-14, American Soc. For Testing of Metals; 2014.

26. AMS 4999A, SAE International; September 2011.

27. Frazier WE, Polakovics D, Koegel W. Qualifying of metallic materials and structures for aerospace applications. *JOM* 2001;**53**:16–18.

28. Slotwinski J, Moylan S Applicability of existing materials testing standards for additive manufacturing materials, Ref NIST.IR.8005. Available from: http://dx.doi.org/10.6028/NIST.IR.8005; 2014.

29. Moylan S, Slotwinski J, Cooke A, Jurrens K, Donmez MA. An additive manufacturing test artifact. *J Res Natl Inst Stand Technol* 2014;**119**:429–59.

30. Measurement science roadmap for metal-based additive manufacturing. Prepared by Energetics Incorporated, Columbia, Maryland for NIST, Dept. of Commerce from presentations made in the Roadmap Workshop on Measurement Science for Metal-Based Additive Manufacturing held on December 4–5, 2012 at the NIST campus in Gaithersburg, MD.

31. Bauza MB, Moylan SP, Panas RM, Burke SC, Martz HE, Taylor JS, et al., 2014 Spring Topical Meeting, vol. 57. p. 86–91.

CHAPTER 4

Microstructure and Mechanical Properties

ABBREVIATIONS AND GLOSSARY

AM	additive manufacturing
DMD	direct metal deposition
DMLS	direct metal laser sintering
EADS	European Aeronautical Defense and Space Company
EBM	electron beam melting
HIP	hot isostatic press
HT	heat treatment
LENS	laser energy net shaping
SAE	Society of Aeronautical Engineers
SR	slot repaired (material)
NR	none repaired (material)
VR	V-groove repaired (material)
WAAM	wire arc additive manufacturing

4.1 INTRODUCTION

AM processes involve layer-by-layer manufacturing that inherently leads to multiple heating and cooling cycles for the bottom layers. This in turn means remelting of the previous layer(s), and reheating and phase transformations for layers that are below. As titanium alloy processing include not only liquid to solid (β) transformation, but, also from BCC-β to HCP-α transformation upon further cooling, microstructure evolution during AM processing is really complex and is affected by any variation of the process parameters that affect the cooling rate and reheating.

4.2 MICROSTRUCTURES

The bulk of the AM work on titanium alloys has been focused on the work horse Ti–6Al–4V alloy. These alloys are well known for their tendency to grow epitaxially and AM processes due to their nature of

Additive Manufacturing of Titanium Alloys. DOI: http://dx.doi.org/10.1016/B978-0-12-804782-8.00004-5

Figure 4.1 Macrostructure showing columnar prior-β grains along the build direction.

directed local heat source aids in promoting epitaxial growth during part build-up. This is well manifested in Fig. 4.1 that shows macrograph of Ti−6Al−4V samples taken from planes parallel to heatflow (laser) direction.[1] It was observed that the prior-β grains in the deposit are columnar in nature, oriented nearly perpendicular to the substrate along the build direction (*z-direction*), and slightly tilted in the direction of laser motion in the $x-z$ or $y-z$ plane.[2] The prior-β grains also grow across multiple deposited layers. Initial solidification of the columnar grains occurs epitaxially from the grains in the base metal, or previously deposited layers, due to similarities in the composition and surface energies of the metal. Subsequent α' transformation has a relationship with prior β matrix. Similar epitaxial growth has been reported in SMD process as well.[3]

Kobryn and Semiatin explored the effect of various process parameters on microstructure evolution and created a solidification map for Ti−6Al−4V alloys produced using DED processes (Fig. 4.2).[4] It is clear that typical processing window during DED processes involving two different laser heat sources (Nd-YAG laser and CO_2

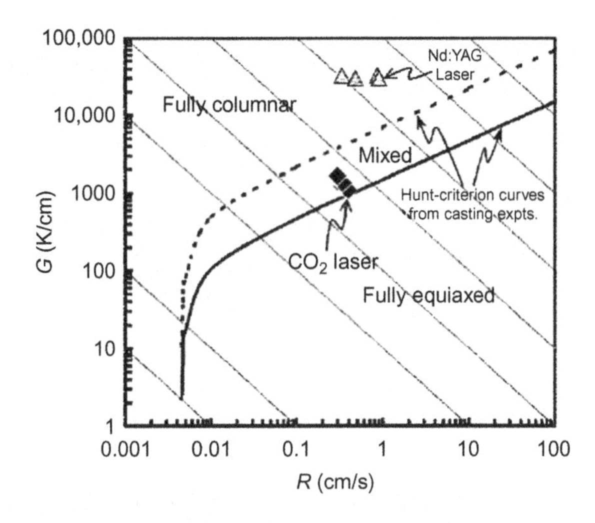

Figure 4.2 Solidification map for DED deposited Ti–6Al–4V alloy.

laser) are both in a regime of temperature gradient–velocity combination that favors columnar (epitaxial) growth in this material. High power density and focused heat source from lasers create a steep temperature gradient during AM processing and aids in columnar growth of these alloys.

When compared to laser processed DED Ti–6Al–4V alloys, microstructures of electron beam processed Ti–6Al–4V alloys contain α–β microstructure. This is a direct consequence of high substrate temperature (approximately 600°C) and slower cooling rate in vacuum environment of EBM process.

As discussed in Chapter 3, Additive Manufacturing Technology, many applications involving AM of Ti-alloys require a post buildup hot isostatic pressing (HIP) operation (refer to ASTM F2924-14, F3001-14 and/or AMS 4999A).[5] Fig. 4.3 shows microstructure of as-built material using DMD process and after subsequent HIP usage and aging. The as-built microstructure shows the typical martensitic α' structure expected for Ti–6Al–4V cooled rapidly from the β phase field, while the HIP and aged material shows the expected grain boundary β and inter-granular coarse α plates. This microstructural transition from as deposited to HIP-aged condition is also reflected through their tensile properties. While tensile strength and yield strength is a little lower after HIP and aging, ductility improves

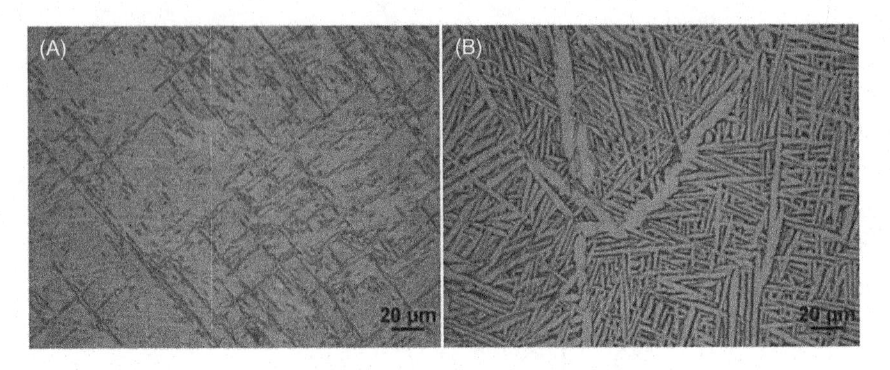

Figure 4.3 Microstructure of DMD built Ti–6Al–4V: (A) as deposited condition and (B) use of HIP and aged as per AMS 4999A. Source: Courtesy of DM3D Technology.

Table 4.1 Tensile Behavior of DMD Built Ti–6Al–4V Material Along Build Direction and Normal to Build Direction

Test Direction	Yield Strength (Mpa)	UTS (Mpa)	Elongation (%)
X–Y build plane	881	971	15.7
Along *Z*-direction (normal to *X–Y* plane)	864	950	14.4

significantly (Table 4.1) as a result of the microstructure changing from martensitic to transformed-β (precipitated-α) structure. Another benefit of HIP is breaking down of the columnar solidification structure and evolution of equiaxed microstructure leading to enhanced isotropic behavior of the material between build direction and its normal (Table 4.1).

4.3 MECHANICAL PROPERTIES FROM VARIOUS AM TECHNOLOGIES

Tensile properties of Ti–6Al–4V fabricated by a number of AM techniques are shown in Fig. 4.4. All of the processes show strength levels superior or comparable to conventional material (cast, forged, and wrought-annealed). As-built materials in laser based processes, such as DMD, LENS and DMLS, exhibit less ductility due to the formation of the martensite α′-phase, however, the ductility can be improved through subsequent HIP and/or heat treatment operation. As a result of reduced residual stress and α-β microstructure, EBM processed Ti–6Al–4V shows greater ductility when compared to laser processed Ti–6Al–4V.

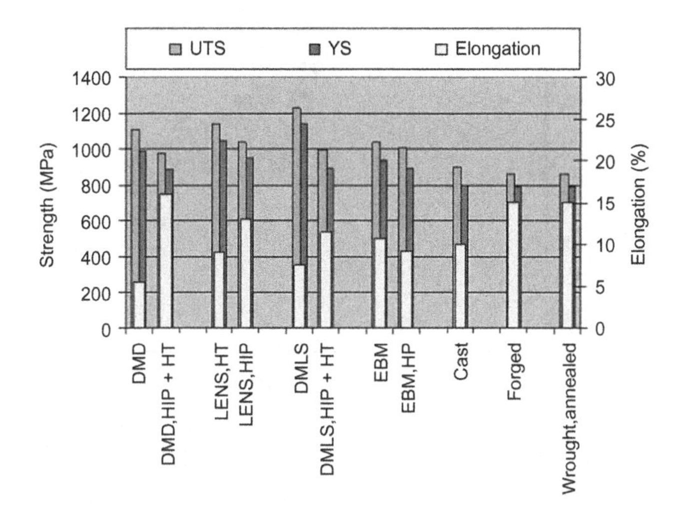

Figure 4.4 Tensile strength, yield strength, and elongation of Ti–6Al–4V alloy built using various AM processes. DMD*: direct metal deposition,*[1] LENS*: laser engineered net shaping,*[6] DMLS*: direct metal laser sintering,*[7] EBM*: electron beam melting,*[7] HIP*: hot isostatic pressing,* HT*: heat treatment.*

Limited work has been performed in other Ti-alloy systems, including high strength α–β alloys, such as Ti-6Al-2Sn-4Zr-6Mo[8] and Ti-6Al-2Sn-4Zr-2Mo[9] and commercial pure Titanium (CP Ti).[1] It appears that tensile strength of these alloys in as deposited condition can be as high, or higher, than wrought properties, while ductility is lower. Another work has defined the effect of interlayer dwell time on residual stress and distortion.[10]

Fatigue properties of AM fabricated and machined Ti–6Al–4V have been tested under high cycle loading conditions and compared with conventionally manufactured Ti–6Al–4V (Fig. 4.5). In general, as built Ti–6Al–4V offers fatigue resistance similar to cast and wrought material, even without a HIP treatment. Similar results have been observed in Ti–6Al–2Sn–4Zr–2Mo alloys where high cycle fatigue behavior of as-deposited alloy is reported to be better than wrought material.[9] It is important to note that these samples were built and machined to specifications for fatigue testing. In contrast to this, when AM built Ti–6Al–4V was tested without surface machining, fatigue life was clearly lower as compared to machined AM Ti–6Al–4V as well as conventional wrought Ti–6Al–4V (Fig. 4.6). This is a clear manifestation of the fact that fatigue is a surface critical phenomenon and rough surface finish of as built AM Ti–6Al–4V negatively impacts fatigue life. Note that the rougher surface finish of

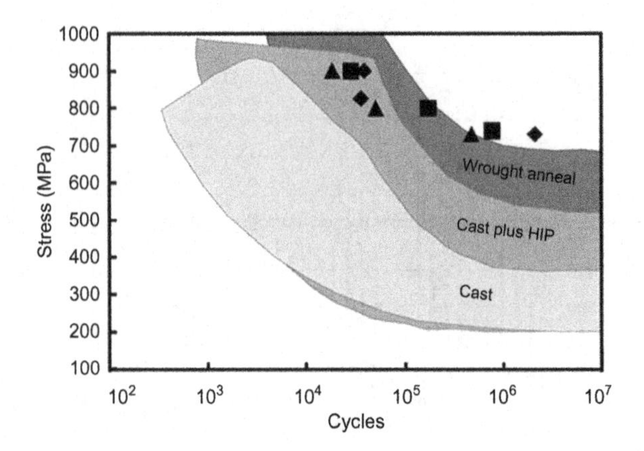

Figure 4.5 Comparison of room temperature fatigue properties of AM fabricated Ti–6Al–4V and conventionally fabricated Ti–6Al–4V. ■, ◆ and ▲ represent properties in the three orthogonal directions: x, y, *and* z, *respectively.* Source: Titanium Development Association, Dayton, Ohio (now The International Titanium Association, http://www.titanium.org).

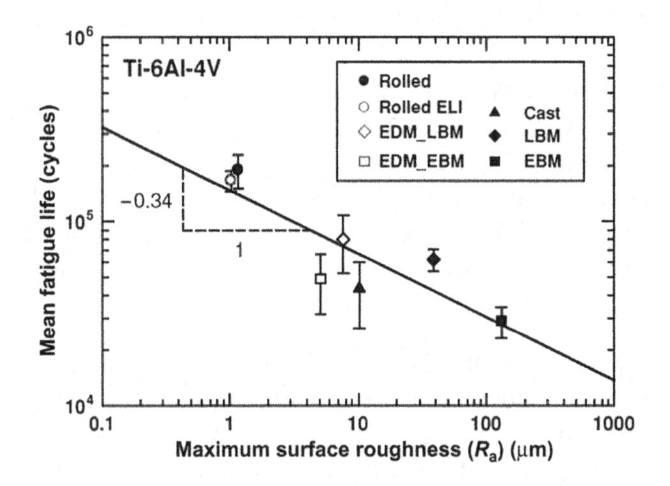

Figure 4.6 Effect of surface roughness on fatigue life of AM manufactured Ti–6Al–4V alloy.[11]

EBM titanium results in lower fatigue life compared to DMLS-built titanium. These results indicate that serious consideration should be given to fatigue behavior if AM parts are used in service without surface finishing and it may not be feasible to use them in fatigue critical applications without surface finishing operations.[11]

DED technologies offer the benefit of repair and remanufacture of damaged parts as well as hybrid manufacturing where features are added on existing parts and preforms (such as simple shape castings,

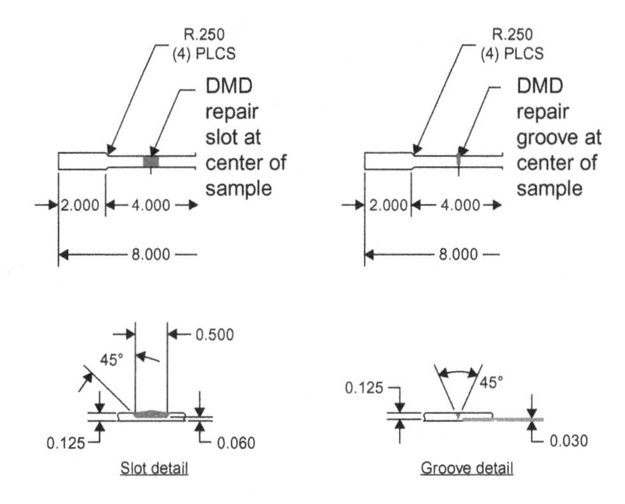

Figure 4.7 Geometry of DMD repaired Ti–6Al–4V samples for tensile testing. Samples were undercut (a slot in one and a V groove in the other) and filled with Ti–6Al–4V using DMD process to simulate property of DMD repaired materials.

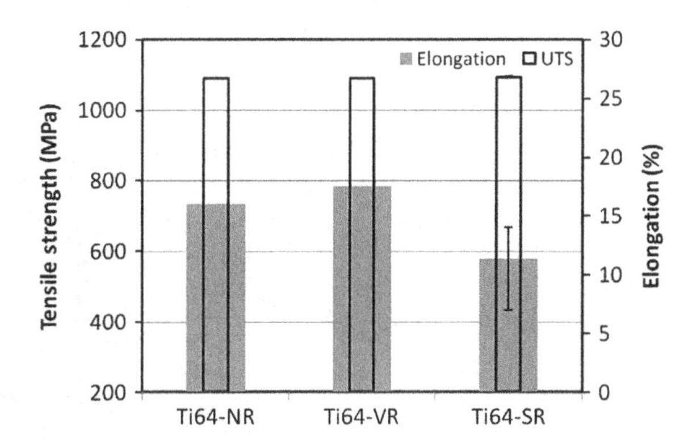

Figure 4.8 Mechanical behavior of DMD repaired Ti–6Al–4V alloy. NR respresents nonrepaired material, while VR is a "V" groove repaired material and "SR" is slot repaired material (refer to Fig. 4.7 for repair geometry).[1]

extrusions, forgings, etc.). Therefore, it is of considerable interest to examine properties of such repairs and feature additions. Fig. 4.7 shows geometry of Ti–6Al–4V specimens that were undercut and grooves filled using DMD process in the center of the gage length of tensile samples. "VR" represents repair of a 45° V-groove 1.5 mm wide and "SR" represents repair of a 12.5 mm area in the center of the specimen. Fig. 4.8 shows that "V" groove repaired and "slot" repaired Ti–6Al–4V samples exhibit comparable tensile strength with nonrepaired ("NR") wrought Ti–6Al–4V sample. This tensile data demonstrates mechanical behavior of DMD repaired material and represent repairs done in the seal areas of jet engine components, such as casings and housings.[3] Ti–6Al–2Sn–4Zr–2Mo samples repaired using laser cladding has shown enhanced high cycle fatigue strength than wrought Ti–6Al–2Sn–4Zr–2Mo material.[8]

4.4 MECHANICAL PROPERTIES AND MICROSTRUCTURE CORRELATION

Fig. 4.9 shows typical microstructures of as built Ti–6Al–4V alloy from various AM processes. Corresponding, tensile properties (UTS and YS) are also plotted as a function of elongation to show the effect of microstructure on mechanical properties. Clearly, laser-based technologies offer higher strength and lower ductility due to the formation of α'-martensite as a result of fast cooling. Electron beam-processed materials exhibit α-β microstructure due to slower cooling in vacuum atmosphere and hot substrate, and results in lower tensile strength and higher ductility. In comparison, microstructure morphology is coarser

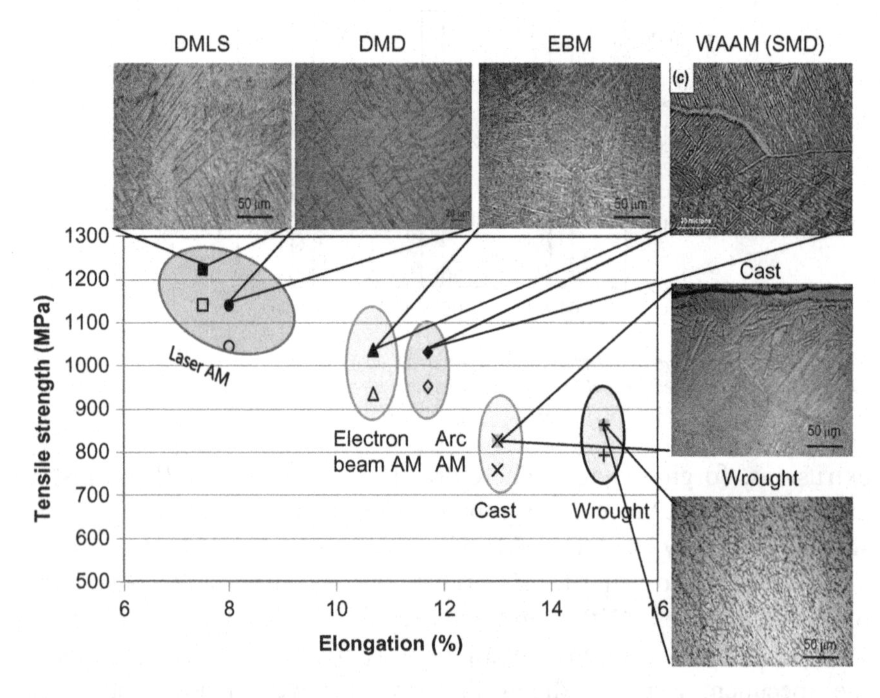

Figure 4.9 Room temperature tensile behavior of AM Ti–6Al–4V alloy produced using various AM technologies and their comparison with cast and wrought material properties.[1,7,12–14] Typical microstructures are also included for comparison. Closed and open symbols represent UTS and yield strength respectively.

in the as-cast material and wrought material has equiaxed α−β microstructure. Arc processed material (WAAM) offers a microstructure similar to cast structure, however, finer in length scale. The effect of these various microstructures is well demonstrated in their tensile behavior.

The relationship of microstructural anisotropy and crystallographic texture with build direction and their effect on tensile behavior has been studied by various workers. While strong orientation of prior-β grains along build direction is observed in all AM processed Ti−6Al−4V material, the relationship between build direction and crystallographic texture is not very well established. Some workers report that β grains grow in [100] direction along build direction and opposite to the heat transfer direction. They further observed that the directional solidification into the β-phase and the subsequent transformation into the α-phase following the Burgers relationship results into an α-texture, where the hcp pole figures look similar to bcc pole figures.[3] Others have reported no clear crystallographic texture after transformation in to α and α′-phase.[15,16] Lower tensile ductility has been observed when samples were tested along *X* or *Y* direction.[15,17] Reduced ductility has been attributed to the less ductile hcp α-phase present along prior β-grain boundaries and normal to tensile axis for these samples.

REFERENCES

1. Dutta B. Private Communication, DM3D Technology; December 2015.
2. Kelly SM, Kampe SL. *Metall Mater Trans A* June 2004;**35A** 1861.
3. Baufeld B, Van Der Biest O, Dillien S. *Metall Mater Trans A* August 2010;**41A**:1917−27. Available from: http://dx.doi.org/10.1007/s11661-010-0255-x
4. Kobryn PA, Semiatin SL. JOM; September 2001. p. 40−2.
5. Titanium alloy direct deposited products Ti−6Al−4V annealed. SAE Aerospace Material Specification (AMS) 4999A, September 2009. <http://www.sae.org/technical/standards/AMS4999A>.
6. <http://www.optomec.com/Additive-Manufacturing-Technology/Laser-Additive-Manufacturing> [accessed July 2013].
7. <http://www.morristech.com/Docs/Ti64ELI%20DataSheet.pdf> [accessed February 2013].
8. Blackwell PL, Wisbe A. *J Mater Process Technol* 2005;**170**:268−76. Available from: http://dx.doi.org/10.1016/j.jmatprotec.2005.05.014.
9. Richter K-H, Orban S, Nowotny S. In: Proceedings of the 23rd International Congress on Applications of Lasers and Electro-Optics; 2004. p. 1−10.

10. Denlingera ER, Heigelb JC, Michalerisb P, Palmer TA. *J Mater Process Technol* 2015;**215**:123–31.
11. Chan KS, Koike M, Mason RL, Okabe T. *Metall Mater Trans A* October 2012. Available from: http://dx.doi.org/10.1007/s11661-012-1470-4.
12. Wang F, Williams S, Colegrove P, Antonysamy A. Microstructure and mechanical properties of wire and arc additive manufactured Ti–6Al–4V. *Metall Mater Trans A* September 2012. Available from: http://dx.doi.org/10.1007/s11661-012-1444-6.
13. Koike M, Greer P, Owen K, Lilly G, Murr LE, Gaytan SM, et al. Evaluation of titanium alloys fabricated using rapid prototyping technologies—electron beam melting and laser beam melting. *Materials* 2011;**4**:1776–92. Available from: http://dx.doi.org/10.3390/ma4101776.
14. ASM Handbook, Vol. 2. Properties and selection: nonferrous alloys and special purpose materials. p. 621, 637.
15. Carroll BE, Palmer TA, Beese AM. *Acta Mater* 2015;**87**:309–20. Available from: http://dx.doi.org/10.1016/j.actamat.2014.12.054.
16. Clark D, Whittaker MT, Bache MR. *Metall Mater Trans B* April 2012;**43B**:388–96.
17. Baufeld B, Van der Biest O, Gault R. *Mater Des* 2010;**31**:S106–11. Available from: http://dx.doi.org/10.1016/j.matdes.2009.11.032.

CHAPTER 5

Comparison of Titanium AM Technologies

ABBREVIATIONS AND GLOSSARY

AM	additive manufacturing
CAD	computer aided design
3D	three dimensions
DED	direct energy deposition
DM	direct manufacturing
DMD	direct metal deposition
DMLS	direct metal laser sintering
EBM	electron beam melting
ELI	extra low interstitial
HAZ	heat affected zone
HIP	hot isostatic press
ID	internal dimension
LENS	laser energy net shaping
MDF	manufacturing demonstration facility
ONR	Office of Naval Research
ORNL	Oak Ridge National Laboratory
PBF	powder bed fusion
SLM	selective laser melting
SLS	selective laser sintering
UAM	ultrasonic additive manufacturing

5.1 TECHNOLOGY COMPARISON

While powder bed fusion (PBF) technologies are suitable for smaller, complex geometries, with hollow unsupported passages/structures, directed energy deposition (DED) is better suited for larger parts with coarser features requiring higher deposition rates. Usage of finer powder grains combined with smaller laser/electron beam size leads to a superior surface finish on the as-built parts from the PBF technologies as compared to DED technologies. However, the majority of AM production parts need some form of finish machining for most

Additive Manufacturing of Titanium Alloys. DOI: http://dx.doi.org/10.1016/B978-0-12-804782-8.00005-7

applications. The ability of the directed energy technologies to add metal onto existing parts allows them to apply surface protective coatings, remanufacture and repair damaged parts, and reconfigure or add features to existing parts, as well as building new parts.

Table 5.1 provides a comparison of capabilities, benefits, and limitations of various AM technologies that are used for producing titanium parts today.

Table 5.1 Comparison of Various Technologies[1–4]

Item	Laser Based PBF (eg, DMLS)	Electron Beam Based PBF (eg, EBM)	Laser Based Directed Energy Deposition (eg, DMD)
Build envelop	Limited	Limited	Large and flexible
Beam size	Small, 0.1–0.5 mm	Small, 0.2–1 mm	Large, can vary from 2 to 4 mm
Layer thickness	Small, 50–100 μm	Small, 100 μm	Large, 500–1000 μm
Build rate	Low, cc/h	Low, 55–80 cc/h	High, 16–320 cc/h
Surface finish	Very good, Ra 9/12 μm, Rz 35/40 μm	Good, Ra 25/35 μm	Coarse, Ra 20–50 μm, Rz 150–300 μm, depends on beam size
Residual stress	High	Minimal	High
Heat treatment	Stress relief required, HIP preferred	Stress relief not required, HIP may or may not be performed	Stress relief required, HIP preferred
Chemistry	ELI grade possible, negligible loss of elements	ELI grade possible, loss of Al needs to be compensated in powder chemistry	ELI grade possible, negligible loss of elements
Build capability	Complex geometry possible with very high resolution. Capable of building hollow channels	Complex geometry possible with good resolution. Capable of building hollow channels	Relatively simpler geometry with less resolution. Limited capability for hollow channels, etc.
Repair/ remanufacture	Possible only in limited applications (requires horizontal plane to begin remanufacturing)	Not possible	Possible (capable of adding metal on 3D surfaces under 5 + 1-axis configuration making repair solutions attractive)
Feature/metal addition on existing parts	Not possible	Not possible	Possible. Depending on dimensions ID cladding is also possible
Multimaterial build or hard coating	Not possible	Not possible	Possible

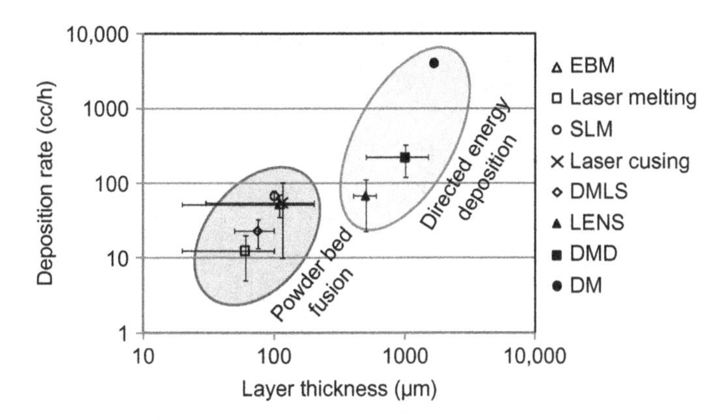

Figure 5.1 Comparison of PBF and DED technologies in terms of layer thickness and deposition rate.[1,5–10]

Fig. 5.1 shows a comparison of PBF technologies against DED technologies in terms of deposition rate and surface roughness. The layer thickness has been used as a measure of roughness here as this determines the roughness of the vertical walls of the structure being built. Clearly, the PBF technologies offer better surface finish as these use a smaller beam size (for both laser and electron beams) and smaller layer thickness when compared to DED technologies; however, as a consequence the deposition rate is also lower for these technologies. Therefore, PBF is more suitable for more accurate, complex small sized objects, while DED is more suitable for building relatively larger parts at a high processing rate, but with a coarser finish surface.

5.2 FREE FORM CAPABILITY

Small beam size and low layer thickness, along with support of the powder bed, allow PBF technologies, such as EBM, DMLS or SLS to produce complex geometries with high precision and unsupported structures. Fig. 5.2 shows one such example of a hydraulic manifold mount for an underwater manipulator built using EBM technology. Building the integrated mount and manifold with internal passageways in a single operation eliminates multiple part fabrication and results in significant cost savings. Good surface finish of the part eliminated finish machining needs on all surfaces, except seal surfaces and threading of screw holes. Generally the PBF technique gives a better surface finish than the DED approach, however for demanding applications

Figure 5.2 Hydraulic manifold built using EBM technology. The part was built at the Manufacturing Demonstration Facility (MDF) in Oakridge National Laboratory through an ONR sponsored project. Source: Courtesy of ORNL, TN.

(such as aerospace) finish machining and/or other surface finishing operations are still required.[11]

5.3 REMANUFACTURING

One of the best application areas suited for DED techniques is the remanufacture and repair of damaged, worn-out, or corroded parts. Due to their ability to add metal on selected locations on 3D surfaces, these technologies can be used to rebuild lost material on various components.[12–14] Closed loop technologies such as DMD offer the particular benefit of minimum heat affected zone (HAZ) in the repaired part and helps to retain the integrity of the part. The close loop control allows DMD to repair parts with a short HAZ and produce a high quality repaired part. Fig. 5.3 shows the cross-section microstructures of the DMD area of a remanufactured turbine blade. The excellent process control during DMD leads to a fully dense microstructure as observed in the vertical cross-section. A layer thickness of about 0.1–0.2 mm has been applied in this case and a minimal HAZ is observed in the as-deposited blade. The DMD vision system plays a significant role in this type of remanufacturing. A calibrated vision system integrated with the machine allows automatic identification of part location in the machine coordinate system and resulting precision processing. Other titanium components that can be repaired include housings, bearings, casing flanges, seals, and landing gears.

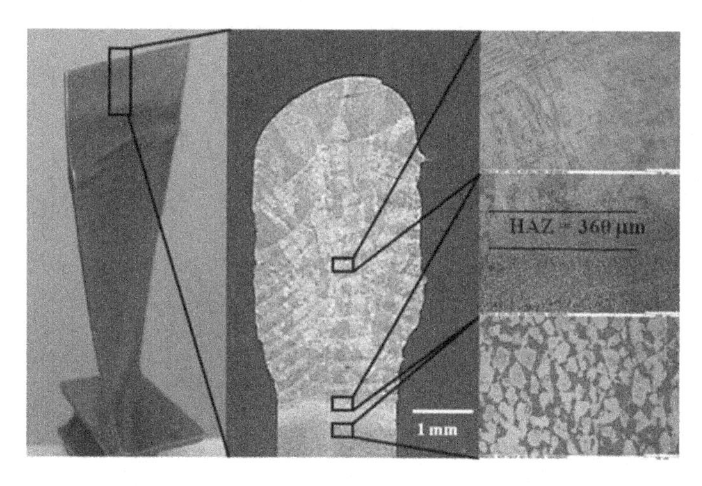

Figure 5.3 DMD repair of turbine components; left*: repaired vane,* middle*: macro cross-section, and* right*: microstructures (top to bottom shows the clad, interface, and base material).* Source: Courtesy of DM3D Technology.

5.4 HYBRID MANUFACTURING FOR LARGE PART AM

Directed energy deposition technologies such as DMD and/or LENS have the ability to add metal on 3D surfaces and thus, allow for the addition of features onto existing parts and/or blanks. This is not possible with the PBF approach. Adding features to a forged or cast preform as opposed to machining of such features can provide the most cost-effective manufacturing option, where a significant reduction of the preform size and weight can be effected through the elimination of the need for a machining allowance. Examples are various casings and housings in jet engines where flanges, bosses, etc. can be added on cast or forged cylindrical performs. This is demonstrated for a feature addition on a titanium fan casing for an aero-engine (Fig. 5.4).

5.5 BI-STRUCTURAL (LESS THAN 100% DENSE STRUCTURES) CONCEPTS

One of the distinct advantages of AM over many conventional manufacturing techniques is its control over the build-up process and its ability to build each layer from a predetermined design. This has allowed usage of AM in the medical implant industry with controlled porosity content in order to facilitate bone growth into the implant.[15,16] Because of their smaller beam size and ability to build unsupported structures, PBF systems are better suitable to fabricate

Figure 5.4 Fan case produced by adding features with AM (laser aided directed energy deposition) to a forged perform. Source: Courtesy of Jim Sears.

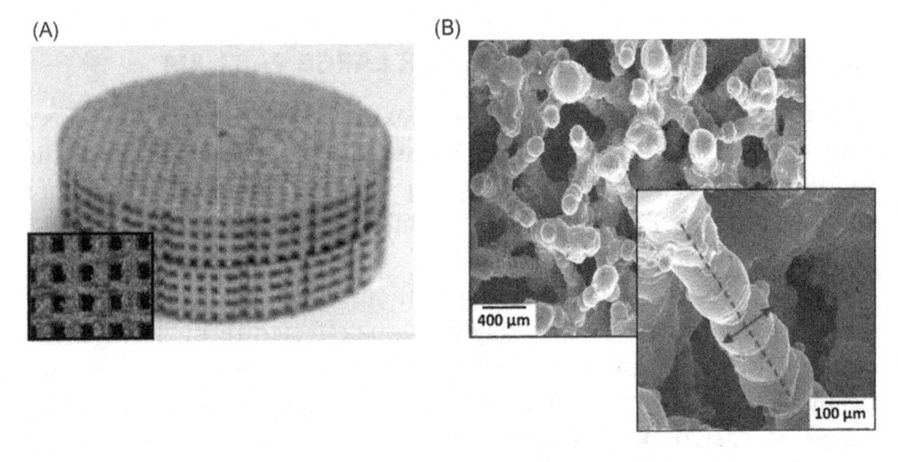

Figure 5.5 (A) Ti–6Al–4V scaffold[18] and (B) porous Ti–6Al–4V[19] for medical implant applications fabricated using SLM technology.

components with a small porosity size and high content and/or with graded porosity content as required by the application. Scaffolds with a porosity content up to about 80% and pore size of about 700 μm[17,18] and random porous structures[19] with porosity content up to about 70% and pore size 300–500 μm have been reported using EBM, SLM and/or DMLS technologies using Ti–6Al–4V material. The LENS technology has been reported to be used to fabricate titanium structures with graded porosity (23–32%) and corresponding tailored elastic modulus (7–60 GPa) in order to match that of the human cortical bone[20] (see also Fig. 5.5).

5.6 BIMATERIAL OR MULTIMATERIAL MANUFACTURING USING AM

One attribute of AM technologies is its ability to fabricate single components with multiple different materials to perform various functions as required by the application. As AM involves layering of materials, one layer at a time, it also allows the introduction of various materials at different layers or in the same layer at different locations. This gives AM a clear edge over all conventional manufacturing processes and provides superiority to AM components.

DED technologies are best suited for fabricating multimaterial components, while other AM technologies, such as ultrasonic AM can also be used for multimaterial component fabrication. Laser based AM has been used to create functionally graded coatings of Rene88DT (Ni-base superalloy) on the Ti–6Al–4V alloy.[21] A controlled experiment produced a continuous gradation of Rene88DT from 0% to 38% Rene88DT alloying over a distance of about 40 mm. This raised the hardness from 450 HV to 750 HV and transformed the microstructure from columnar α(Ti) + β(Ti) to α(Ti) + β(Ti) + Ti_2Ni to equiaxed β(Ti) + Ti_2Ni (Fig. 5.6). Such precise engineering of microstructures and properties can enhance performance of components drastically. Similarly, graded alloying has been performed with Ti–Mo alloys and Ti–V alloys using other DED technologies.[22]

Multilayered hybrid metal laminates have been studied widely for armor applications as gradients of different metals can be used to design highly customized through thickness mechanical properties. Through thickness properties such as strength, toughness and stiffness can be varied to produce a system with the highest performance at the lowest weight. Ultrasonic additive manufacturing (sheet lamination) has been used to produce armor panels with a combination of aluminum and titanium alloys (Fig. 5.7).[23] Typically layers are in the range of 150 microns and each layer can be changed to build a gradient structure through the thickness.

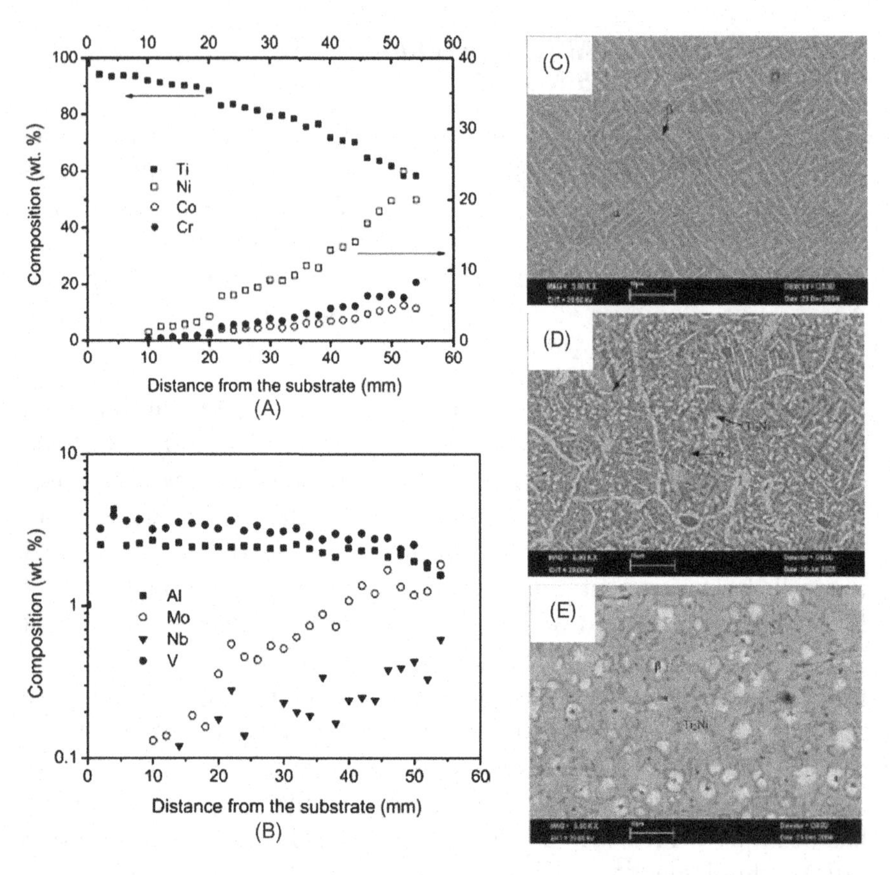

Figure 5.6 Functionally graded coating of Rene88DT alloy on Ti–6Al–4V using laser based AM (DED). (A) and (B) show composition gradients of various elements along the depth from the top surface. The corresponding microstructures are shown in (C) Ti–6Al–4V–0%Rene88DT, (D) Ti–6Al–4V–19%Rene88DT, and (E) Ti–6Al–4V–38% Rene88DT.[21]

Figure 5.7 Laminated armor after ballistic testing. The part was built using ultrasonic additive manufacturing (UAM). Source: Courtesy of Mark Norfolk, Fabrisonic LLC.

REFERENCES

1. <http://www.eos.info/additive_manufacturing/for_technology_interested> [accessed July 2013].
2. <http://www.arcam.com/technology/additive-manufacturing/> [accessed July 2013].
3. <http://www.dm3dtech.com/index.php/expertise-innovations/experticeandinnovations-dmddtechnology> [accessed July 2013].
4. <http://www.morristech.com/Docs/Ti64ELI%20DataSheet.pdf> [accessed February 2013].
5. Dutta B. Private Communication. DM3D Technology; July 2013.
6. Stecker S, Lachenberg KW, Wang H, Salo RC. Advanced electron beam free form fabrication methods & technology. In: AWS conference; 2006. p. 35–46.
7. <http://www.optomec.com/Additive-Manufacturing-Technology/Laser-Additive-Manufacturing> [accessed July 2013].
8. <http://resources.renishaw.com/en/details/brochure-the-power-of-additive-manufacturing--57719> [accessed July 2013].
9. <http://stage.slm-solutions.com/index.php?slm-500_en> [accessed November 2013].
10. <http://www.industriallaser.com.au/pdf/X%20Line%201000R%20Brochure.pdf> [accessed November 2013].
11. Dehoff R, Duty C, Peter W, Yamamoto Y, Chen W, Blue C, et al. Case study: additive manufacturing of aerospace brackets. *Adv Mater Process* 2013;**171**(3):19–22.
12. Dutta B, Palaniswamy S, Choi J, Song LJ, Mazumder J. Additive manufacturing by direct metal deposition. *Adv Mater Process* May 2011;33–6.
13. Dutta B,Palaniswami S, Choi J, Mazumder J. Rapid manufacturing and remanufacturing of DoD components using direct metal deposition. AMMTIAC Quarterly 6(2), 5–9.
14. Dutta B, Natu H, Mazumder J. Near net shape repair and remanufacturing of high value components using DMD. In: TMS proceedings, vol.1: Fabrication, materials, processing and properties; 2009. p. 131–8.
15. Sidambe AT. *Materials* 2014;**7**:8168–88.
16. Wally ZJ, van Grunsven W, Claeyssens F, Goodall R, Reilly GC. *Metals* 2015;**5**:1902–20.
17. Wieding J, Jonitz A, Bader R. *Materials* 2012;**5**:1336–47.
18. Jonitz-Heincke A, Wieding J, Schulze C, Hansmann D, Bader R. *Materials* 2013;**6**:5398–409.
19. Kima TB, Yueb S, Zhanga Z, Jonesc E, Jonesa JR, Lee PD. *J Mater Process Technol* 2014;**214**:2706–15.
20. Bandyopadhyay A, Espana F, Balla VK, Bose S, Ohgami Y, Davies NM. *Acta Biomater* 2010;**6**:1640–8.
21. Lin X, Yue TM, Yang HO, Huang WD. *Metall Mater Trans A* January 2007;**38A**:127–37.
22. Collins PC, Banerjee R, Banerjee S, Fraser HL. *Mater Sci Eng* 2003;**A352**:118–28.
23. Norfolk M. Private Communication. Fabrisonic LLC, Received on December 30, 2015.

CHAPTER 6

Markets, Applications, and Costs

ABBREVIATIONS AND GLOSSARY

AM	additive manufacturing
BALD	bleed air leak detector (bracket)
CMF	implant cranio maxillofacial implant
DED	direct energy deposition
DMLS	direct metal laser sintering
DMD	direct metal deposition
3D	three dimensional
EBM	electron beam melting
FEA	finite element analysis
HAZ	heat affected zone
HPDC	high pressure die casting
JSF	Joint Strike Fighter (F-35)
MRO	maintenance repair and overhaul
NIST	National Institute for Science and Technology
ORNL	Oak Ridge National Laboratory
PBF	powder bed fusion
SLM	selective laser melting
SLS	selective laser sintering
UAM	ultrasonic additive manufacturing
UK	United Kingdom (Great Britain)
USA	United States of America
Z	atomic number of an element

6.1 MARKETS AND APPLICATIONS FOR TI AM

3D printing and additive manufacturing (AM) has been famously touted as a new industrial revolution in *The Economist*.[1] With the maturity of the technology and enhancements in the quality and productivity of AM parts, it is becoming part of mainstream manufacturing. As the overall AM market continues to grow at a rapid rate, the market for titanium AM follows this trend. Analysts project that revenues for titanium

Additive Manufacturing of Titanium Alloys. DOI: http://dx.doi.org/10.1016/B978-0-12-804782-8.00006-9

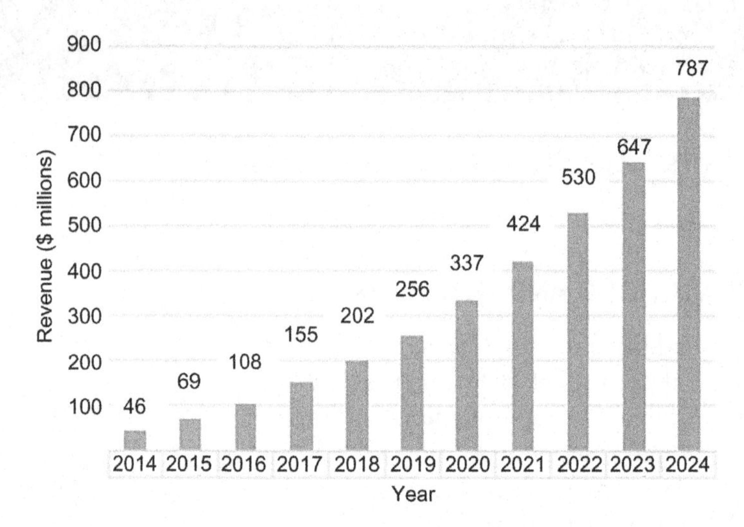

Figure 6.1 Projected revenue for titanium AM powders from 2014 through 2024. Source: SmarTech Publishing LLC.[2]

powders used in AM will reach more than US$330 million by 2020, corresponding to 730,500 kg or 1,610,477 lb (Fig. 6.1).[2] Aerospace and medical industries, being the two largest consumers of titanium metal, are naturally the two largest drivers of titanium AM. Extensive exploration is currently underway for the usage of AM titanium parts in aerospace and medical applications. Other applications for AM include use in chemical, defense, and other industries. While powder bed fusion technologies are suitable for smaller, complex geometries, with hollow unsupported passages/structures, directed energy deposition is better suited for larger parts with coarser features requiring higher deposition rates. Usage of finer powder particles combined with a smaller laser/electron beam size leads to a superior surface finish on the as-built parts from the powder bed fusion technologies as compared to directed energy deposition technologies. However, the majority of the AM parts need finish machining for most of practical applications. The ability of the directed energy technologies to add metal on existing parts allow them to apply surface protective coatings, remanufacture and repair of damaged parts and reconfigure or add features to existing parts, as well as building new parts.

6.1.1 Aerospace Industry

Aerospace and defense are the largest consumers of titanium metal. Titanium applications in the aerospace industry is currently being explored for smaller structures in aircraft engines such as brackets and

Figure 6.2 Titanium pump housing built using SLM process.[3] Source: Courtesy of Richard Grylls, SLM Solutions.

housings, but may expand into larger structural components which will drive demand. By 2020, aerospace is forecast to consume almost 155,000 kg (341,717 lb) of titanium.[1] A majority of small sized part applications is focused on weight savings through innovative designs that result in fuel savings and significant economic benefit during the service life of the aircraft. In addition to this, AM of titanium components is also focusing on parts with high buy-to-fly ratio, to reduce the input weight of titanium. Buy-to-fly ratio is defined as the weight of the originally purchased stock material to the weight of the final finished part. Typical aerospace components can have buy-to-fly ratio of 10:1–20:1, and as high as 40:1 in some cases when fabricated using conventional manufacturing processes. AM can reduce this ratio significantly and make it close to 1:1. This not only saves cost for expensive titanium alloys, but also reduces machining time and cost. The result is a multifold saving for the manufacturing of such components. Larger components are expected to exhibit even more significant savings by reducing the buy-to-fly ratio when fabricated by AM. This will further expand the growth of both titanium metal production as well as AM of titanium alloys. Fig. 6.2 shows a titanium pump housing aerospace applications built using SLM technology.

Another significant opportunity for AM in aerospace industry is in the area of maintenance repair and overhaul (MRO) market.

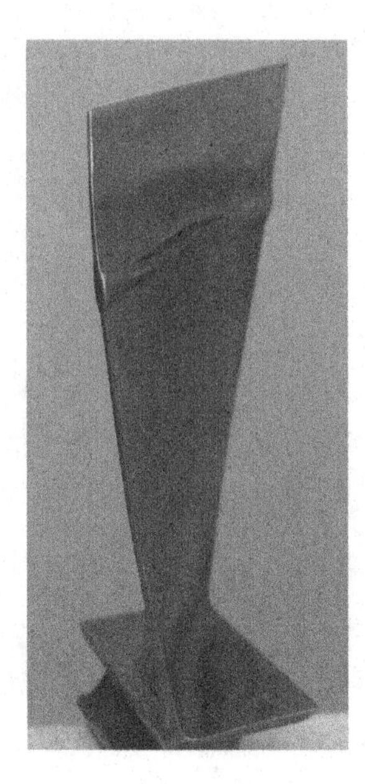

Figure 6.3 Vane tip repair using the DMD process.[4] Source: Courtesy of DM3D Technology.

Sustainability needs and ever-increasing material costs are forcing airline companies to explore repair options that go beyond current repair capabilities. AM offers the benefit of metallurgical bond, a reduced HAZ, reduced distortion and better quality of repaired components, while saving significant cost and reducing downtime. Fig. 6.3 shows a titanium compressor vane repaired using DMD technology.

6.1.2 Medical Industry

Titanium AM has good prospects in the medical markets due to bio-inertness and as-manufactured bone ingrowth performance. Current production of titanium implants using AM is growing rapidly, with new products in spine, hip, knee, and other orthopedic areas. Projections are that medical applications of AM titanium will account for roughly 274,000 kg (604,067 lb) in 2020 due to this growth.[1] Fig. 6.4 shows examples of a CMF implant, an acetabular cup and a tibial implant built using EBM technology.[5,6] These technologies also have the potential of fabricating patient specific custom implants which better suit the needs for this application. Fig. 6.5 shows a Ti–6Al–4V scaffold built using DMD technology.[7]

Figure 6.4 Medical implant applications: (left) CMF implant using EBM technology, (middle) acetabular cup using EBM technology[5]*, (right) tibial implant using EBM technology.*[7] Source: Middle, Courtesy of Bruce Bradshaw, ARCAM.

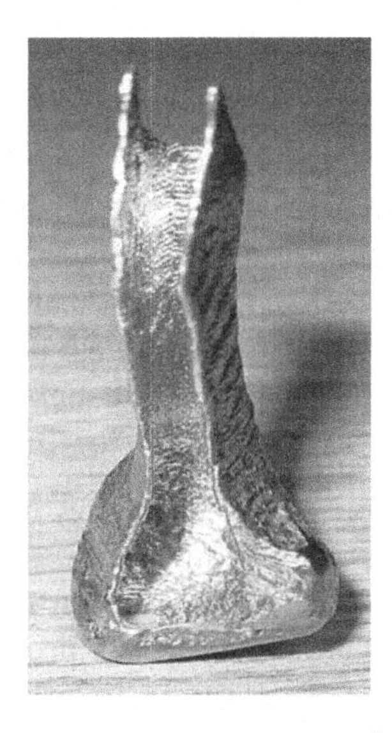

Figure 6.5 Ti–6Al–4V scaffold which was produced by the DMD process.[7]

6.1.3 Other Industries

Other industries, such as chemical, oil and gas, are also exploring titanium AM applications. Below are some examples of multimaterial manufacturing involving titanium.

The ability of a shield to stop radiation generally increases with the atomic number (Z) of the shield material. Graded Z shielding is a combination of several different materials with differing atomic numbers. Graded Z shields typically have much higher performance per unit weight than shields made of a single material. Space structures and satellites have been shielded for years using laminates of several different metals

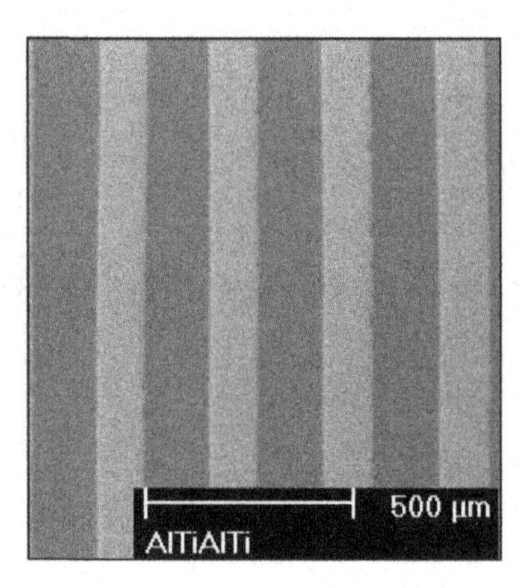

Figure 6.6 Graded layers of Al and Ti material fabricated for radiation shielding application.[8] Source: Courtesy of Mark Norfolk, Fabrisonic LLC.

(Al, Ta, Cu) including titanium. However, these laminates are difficult to produce as fusion based welding processes lead to brittle intermetallics. Ultrasonic additive manufacturing (UAM) can be used to fabricate such graded material shields at room temperature and thus eliminate the concerns of intermetallic formation. Fig. 6.6 shows one such example of a radiation shield fabricated using aluminum and titanium.[8]

Another example of multimaterial manufacturing application is in armor. Multilayered hybrid metal laminates have been studied widely for armor applications as a gradient of different metals can be used to design highly customized structures with attractive 'through thickness' mechanical properties. Through thickness properties such as strength, toughness and stiffness can be varied to produce a system with the highest performance at the lowest weight. UAM has been used to produce armor panels with a combination of aluminum and titanium alloys. Typically layers are in the range of 150 microns and each layer can be changed to build a gradient structure through thickness.[8]

6.2 COST ANALYSIS OF AM PART MANUFACTURING AND COMPARISON WITH CONVENTIONAL METHODS

There have been extensive studies on cost analysis and costing models for AM based on PBF processes. A comprehensive summary of this work can be found in a NIST report.[9] As in any manufacturing process, the cost of AM depends on factors such as machine cost, material cost, build time, energy consumption, labor, and overhead costs including facilities and other costs. While most of these costs are easily understood, part size with respect to the machine build envelop plays a major role in determining per piece build time and cost. This is because PBF processes require filling out the entire build area with powder for each layer, and the more efficiently the build area is filled with parts, the per piece build time is reduced as well as the powder usage. Two popular cost models are:

- Hopkinson and Dickens' model calculates the average cost per part by dividing the total cost by the total number of parts manufactured in a year. Total cost is the sum of machine costs (depreciated over 8 years), labor costs, and material costs. Three additional assumptions are: (1) the system produces a single type of part for 1 year, (2) it utilizes maximum volumes, and (3) the machine operates for 90% of the time.
- Ruffo, Tuck and Hague's model based on particular activities with the machine. They calculate total cost of a build as sum of raw material cost and indirect costs (hourly rates for machine cost, labor cost, administrative cost, and facility cost multiplied with the particular build time). The cost per part is calculated as the total cost of a build divided by the number of parts in the build. If the build consists of parts with different sizes and shapes, per part cost is calculated as volume fraction of the specific part with respect to the total build volume, multiplied with the total build cost.

While the Hopkinson model generates a flat cost for the machine, Ruffo's model takes into account the build envelop usage for a single part, also the volume of parts and volume of various different parts and likely yields a more accurate costing for part production using PBF systems (Fig. 6.7). Both the models are targeted towards production costs and do not consider the engineering cost involved in reengineering the part for the process, programming, as well as any post-processing, etc.

Fig. 6.8 shows a typical cost breakdown of the various steps involved in AM of titanium components using DED technologies. This model is based on the following assumptions: (1) small batch size (between 10 and 30 parts), (2) medium sized part, ~600–900 mm (2–3 ft) in size and relatively simple geometry. Any and all of these factors can significantly influence the cost. The batch size plays a major role in costing for small batch sizes, while part size plays a more significant role in costing for larger batch sizes.

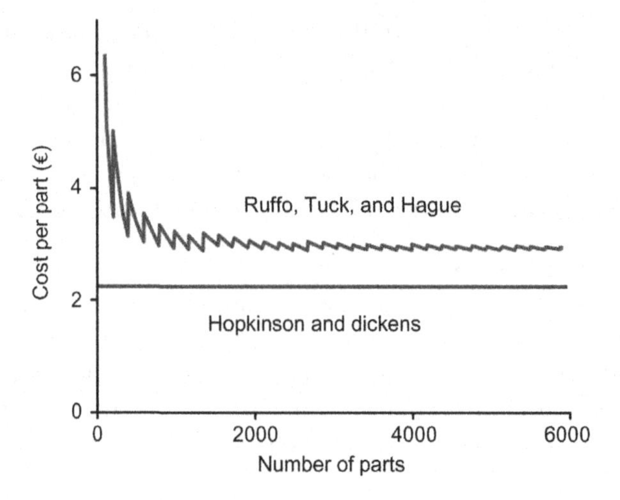

Figure 6.7 Two different cost models for a laser sintering system (PBF process).[9]

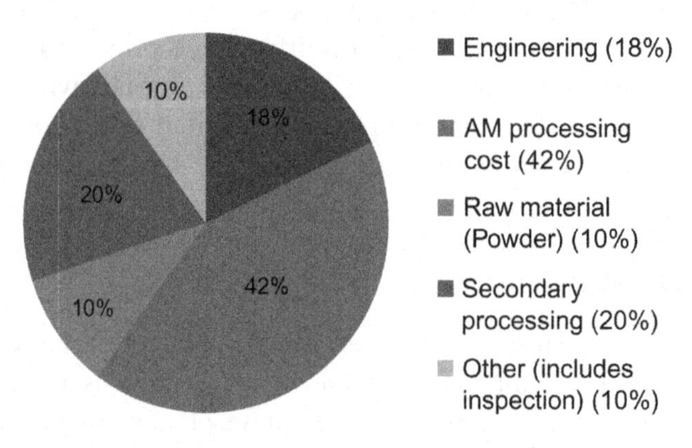

Figure 6.8 Typical cost breakdown of various steps involved in AM of titanium.[4]

6.3 ECONOMICS OF AM

Success in expanding AM in the manufacturing industry depends on selection of the right applications. Appropriate applications for AM include a long lead time, complex components, low volume expensive components, weight savings through innovative designs, cost savings in high buy-to-fly ratio parts, cost-effective remanufacturing, customized medical implants, and multimaterial components. Below are few case studies that expand on these aspects of AM.

Benefiting from tool-less manufacturing makes AM an attractive manufacturing option for small batch sizes when compared with

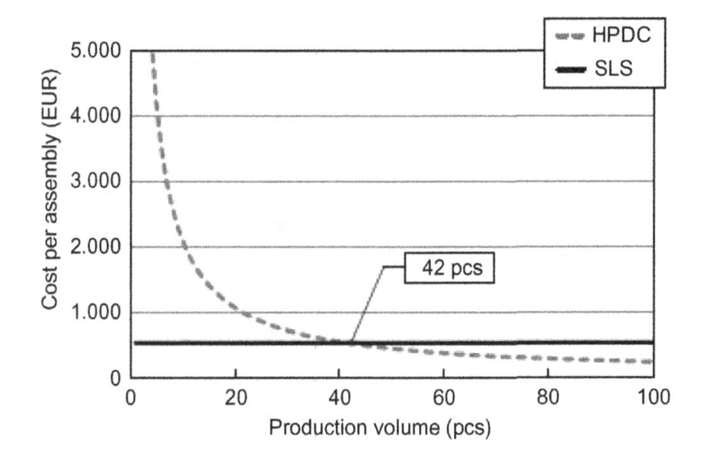

Figure 6.9 Cost comparison for a landing gear made using AM (SLS process) and the conventional HPDC process.[10]

conventional manufacturing techniques, such as casting, forging, and extrusion. Elimination of tooling needs not only reduces per piece cost, but also drastically reduces manufacturing lead time that directly translates into significant economic benefits. However, relatively lower throughput renders AM less attractive for high volume manufacturing. This is illustrated in Fig. 6.9 that compares the per piece cost of manufacturing of a landing gear component using the DMLS process with the high pressure die casting (HPDC) process. As the part quantity exceeds 42 pcs, HPDC becomes a more cost-effective solution, while AM is less costly at low quantities.[10]

Both the aerospace and medical industries use titanium and its alloys extensively for their applications requiring high performance, low weight, bio compatibility, etc. Titanium and its alloys are not only expensive, but also difficult to machine using traditional manufacturing methods resulting in high manufacturing costs. AM offers a unique opportunity by reducing the raw material requirements as well as the extent of required post-machining operations. A cost model of a typical aeroengine titanium part made using the AM process in comparison with machining from a ring forged material stock has been done by Allen.[11] This work concluded that AM is cost effective in instances where the buy-to-fly ratio is 12:1 or more. The buy-to-fly ratio is calculated as the weight of the precursor billet divided by the weight of the final component. It is a means for indicating how much material must

Figure 6.10 BALD bracket for Joint Strike Fighter built using EBM technology. Source: Department of Energy, Oak Ridge National Laboratory; Lockheed Martin.

be machined away. The cost comparisons show that AM is an attractive option for components with a high buy-to-fly ratio, have a complex shape that requires significant machining, has a high material cost, and is difficult to machine.

The above analysis is illustrated through the following example. Researchers at the Oakridge National Laboratory built a Ti–6A–4V Bleed Air Leak Detect (BALD) bracket for the Joint Strike Fighter (JSF) engine using EBM technology (Fig. 6.10).[12] Traditional manufacturing from wrought Ti–6Al–4V plate costs almost $1000/lb due to a high (33:1) buy-to-fly ratio. In contrast this ratio is just over 1:1 ratio for the AM built part. Estimated saving through AM is about 50%.

One of the major benefits of the powder bed fusion processes is the ability to create hollow structures and thereby allow weight savings. The aerospace industry, where weight savings can have a very significant impact, is actively looking into this characteristic of AM processes. A case study involving a seat buckle for a commercial passenger jets is a very good example of this.[13] A lightweight seat buckle with hollow structures was designed based on extensive FEA studies to ensure enough strength against shock loading. The part was produced using DMLS from the Ti–6Al–4V alloy (Fig. 6.11). Replacement of a conventional steel buckle with a hollow AM

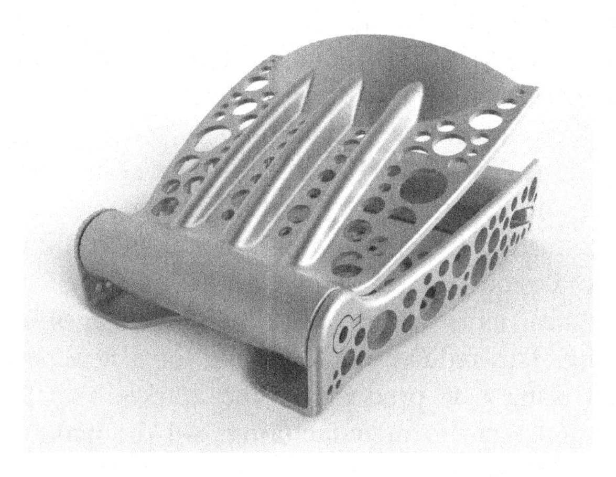

Figure 6.11 Seat buckle produced using DMLS technology.[13] Source: © Mike Ayre, Crucible Design Ltd.

titanium buckle resulted in 85 g weight saving per buckle (55% weight reduction). An Airbus A380 with 853 seats will thus result in a possible weight saving of 72.5 kg. According to the project sponsor, Technology Strategy Board, UK, this weight saving translates to 3.3 million liters of fuel savings over the life of the aircraft that is equivalent to £2 million (US$3 million), while the cost of making all the buckles using DMLS is only £165,000 (US$256,000).

Direct deposition techniques such as DMD can not only be used to create parts, but these technologies can also be used for remanufacturing, repair, and/or feature building on existing parts. Damaged expensive aerospace titanium components, such as bearing housing, flanges, fan blades, casings, vanes, and landing gears can be rebuilt using AM technologies at 20–40% of the cost of the new parts.[4] Worn out flanges in jet engine casings have been rebuilt using DMD at less than half of the cost of a new part. Extensive work is underway to investigate the feasibility of using such technologies to salvage components that are mismachined during conventional manufacturing. Successful realization of these efforts will have a very significant impact on the titanium manufacturing industry. While most of the leading commercial activities in the AM industry is concentrated in the United States and Europe, significant efforts are also underway in many parts of the world, including China.[14]

From a supply chain perspective, AM can have a significant impact and differs significantly from conventional manufacturing.[15] While conventional manufacturing typically relies on a three-tier system (consisting of manufacturer, tooling supplier and warehouse), AM offers the opportunity of a one-tier system where the second tier of tooling suppliers are not needed as AM is tool-less manufacturing. Due to the flexibility of AM technology to switch from one part production to another quickly, AM also eliminates the need for large part warehouse inventories and thus eliminates a third tier in the supply chain. Besides this, AM can be extremely effective for new product launches, as it allows quick design changes and does not require tooling. This reduces the lead time for a new part and allows customers to bring new products to the market very quickly, from concept to engineering to manufacturing. All the makes AM a great tool for lean manufacturing.

6.4 SUSTAINABILITY OF AM

AM impacts all three major areas of sustainability: economics, environment, and social sustainability.[16] Impacts on environmental sustainability arises from energy savings and material savings in AM as discussed in the previous section. All material savings are also indirectly related to energy savings (as it eliminates the energy consumption associated with raw material manufacturing), besides the fact that lightweight AM structures in aerospace and other transportation applications leads to energy savings throughout the life of the component. Remanufacturing through AM also leads to material savings, reduces scrap, and preserves natural resources further aiding environmental sustainability. Social sustainability of AM comes from the improvement in quality of human life and increased consumer satisfaction. For example, Ti-medical applications using AM offer the opportunity of patient specific customization and therefore, enhanced recovery that can lead to very significant impact on quality of human life and satisfaction of the patients. Economic sustainability is an indirect result of the two previous factors, namely, environmental and social sustainability, as well as a result of direct cost reduction achieved in many titanium applications through AM processing.

REFERENCES

1. The Economist, February 10, 2011. <http://www.economist.com/node/18114221/print>.

2. Adv Mater Process, January 2016, p. 6.

3. Grylls R. Private Communication. SLM Solutions; 2016.

4. Private Communication. DM3D Technology; 2016.

5. Bradshaw B. Private Communication. ARCAM; 2016.

6. Murr LE, Gaytan SM, Martinez E, Medina F, Wicker RB. *Int J Biomater* 2012;**1** Article ID 245727, 14 pages.

7. Dinda GP, Song L, Mazumder J. *Metall Mater Trans A* December 2008;**39A**:2914–22.

8. Norfolk M. Private Communication. Fabrisonic.

9. Thomas DS, Gilbert SW. NIST Special Publication 1176, U.S. Department of Commerce, December 2014. Available from: http://dx.doi.org/10.6028/NIST.SP.1176.

10. Atzeni E, Salmi A. *Int. J. Adv. Manuf. Technol.* 2012;**62**:1147–55.

11. J Allen. An investigation into the comparative costs of additive manufacture vs. machine from solid for aero engine parts. In Cost effective manufacture via net-shape processing, 17-1–17-10. Meeting Proceedings RTO-MP-AVT-139. Paper 17. DTIC Document. <http://www.rto.nato.int/abstracts.asp> ; 2006.

12. Dehoff R, Duty C, Peter W, Yamamoto Y, Chen W, Blue C, et al. Case study: additive manufacturing of aerospace brackets. *Adv Mater Process* 2013;**171**(3):19–22.

13. <http://www.manufacturingthefuture.co.uk/_resources/case-studies/TSB-AirlineBuckle.pdf> [accessed February 2016].

14. Qian M. Private Communication; February 2016.

15. Scott A, Harrison TP. *3D Printing* 2015;**2**(2).

16. Garrett White Daniel Lynskey, Session B6 3039, 2013,136.142.82.187.

CHAPTER 7

Recent Developments and Projections for the Future of Titanium AM

ABBREVIATIONS AND GLOSSARY

AM	additive manufacturing
ADMA	advanced materials (corporation)
CAD	computer aided design
CAD/E/M	computer aided design/engineering/manufacturing
CSG-BRep	constructive solid geometry—boundary representation
CT/MRI	computerized tomography/magnetic resonance imaging
CVD	chemical vapor deposition
3D	three dimensions
DFAM	design for additive manufacturing
DMD	direct metal deposition
EBM	electron beam melting
MER	metallurgical and electrolytic research (corporation)
NIST	National Institute for Science and Technology
PREP	plasma rotating electrode process
PTA	plasma transferred arc
PVD	physical vapor deposition
S–N	stress–number of cycles (fatigue test)
UK	United Kingdom
WAAM	wire arc additive manufacturing

7.1 INTRODUCTION

As indicated in Chapter 6, Markets, Applications, and Costs, titanium AM promises very strong growth potential in the coming years. Just the revenue from AM titanium powder production is expected to reach more than $330 million by 2020 and almost $800 million by 2024.[1] The forecast is that with the increased adoption of the technology, the equipment cost will be decreased and advancement of the technology will enhance the process speed significantly (400%) over the next 5 years.[2] Based on these assumptions, titanium AM industry can be

Additive Manufacturing of Titanium Alloys. DOI: http://dx.doi.org/10.1016/B978-0-12-804782-8.00007-0

estimated at US$1 billion by 2020 and US$3 billion by 2024. Machine manufacturing and servicing for titanium AM is also expected to grow in a proportionate manner. AM is a more cost effective way of manufacturing complex titanium components and is therefore expected to expand the titanium industry as a whole.

7.2 NEW DEVELOPMENTS IN AM MATERIALS

With the acceptance of AM in main stream manufacturing, researchers are focusing on exploiting the added benefits of AM: (1) ability to build a single part with multiple materials or graded materials[3–4] and (2) applying wear resistant coatings on titanium parts.[4–6] An example is graded coating of Rene 88 onto Ti–6Al–4V[6] and Mo–WC coating on Ti–6Al–4V material[5] and surface alloying with N, C or B to form hard coatings of TiN, TiC or TiB.[5] Such coatings, when integrally built with 3D titanium components, is expected to offer superior properties compared to traditional titanium parts that are coated with CVD or PVD coatings today. Building multimaterial components requires process control development as well as developing software capability that will allow building a single object with multiple materials directly from the CAD data. Newer alloys such as high Nb containing TiAl alloys are also being pursued.[7]

7.3 RESEARCH AND DEVELOPMENT IN TITANIUM AM

Mathematical modeling and simulation of the AM process as well as development of predictive process–structure–property relationships integrated with CAD/E/M tools is a thrust area in AM research. It is necessary to have computational methods for analyzing materials and material combinations, as well as designing materials and their combinations that can correlate material to processes to structure to property. Understanding the AM process, the thermal history, and microstructure development during AM is key to better and robust process control and obtaining tailored microstructures.[8,9] As AM is a layer-by-layer process it offers the opportunity of altering the process parameters as desired to create designed microstructures in the parts. Work is underway to better understand the effect of changing various process parameters and how that can be implemented in creating parts with designed microstructures in various locations. Such work has already been initiated in Ni-base alloys and can also be continued in titanium AM.[10] Physics-based multiscale predictive models that account for geometric accuracy, spatial material properties, defects, surface characteristics, and other variations are currently not available for AM. A detailed list of various research priority areas can be found in the report from the NIST organized workshop on "Measurement Science Roadmap for Metal-Based Additive Manufacturing".[11]

7.4 NEW DESIGN INITIATIVES

Most of the current AM applications are to replace existing components that have been designed for other traditional (subtractive manufacturing) manufacturing processes. This often results in limiting the benefits of AM for the particular application and sometimes, limits application of AM for the particular component. Integration of the design function—also known as design for additive manufacturing (DFAM)—with optimization software will lead to more predictable processes and higher quality products. In a prior Roadmap workshop organized by National Science Foundation (NSF), DFAM was characterized as: synthesis of shapes, sizes, geometric mesostructures, and material compositions and microstructures to best utilize manufacturing processes capabilities of achieve desired performance and other lifecycle objectives.[12] One such example is the need for a CAD software that will allow for the design of components for a particular AM process while reducing the mass of the component, but meeting required design criterion for the component. Work is being pursued to better integrate CAD software for various AM processes. Designing efficient support structures for overhangs is a critical design area that is gaining a lot of attention.

Another area of research focus is software development for interpretation of the CT/MRI imaging data, and subsequent translation into CAD data to build customized biomedical products, such as patient specific implants for the orthopedic industry.[7] However, commercial CAD systems are ill-suited for modeling parts with complex constructions (eg, lattices or scaffolds) with thousands of shape elements or with material distributions, such as functionally graded materials or tissue constructs. When more than 1000 surfaces or parts are modeled, CAD systems tend to run very slowly and use hundreds of megabytes or several gigabytes of memory due to the nature of parametric CSG-BRep (constructive solid geometry—boundary representation) technology.[12] Similarly, CAD systems have limitations in representing parts with multiple materials and this is a barrier to modeling multimaterial components.

7.5 LOW-COST TITANIUM AM

The vast majority of Ti AM components have been fabricated using high-cost spherical gas atomized or PREP (plasma rotating electrode process) powder (see chapter: Raw Materials for Additive Manufacturing of Titanium). As seen in Chapter 6, Markets, Applications, and Costs (Fig. 6.8), raw material (mostly powder) cost can be almost 10% of the overall component cost. As AM continues to evolve, a focus is on driving the powder cost down to make it a competitive alternative solution. Recent work has demonstrated that Titanium AM parts can be successfully produced using much lower cost angular powders.[13,14] In one program,[13] titanium sponge was blended with Al and V powder or an Al/V master alloy to produce the Ti–6Al–4V composition. After processing this combination through a plasma transferred arc (Fig. 7.1) the as-fabricated tensile properties were at cast and wrought (ingot metallurgy) levels: 980 MPa (142 ksi) UTS, 882 MPa (128 ksi) YS and 10% elongation, and S–N fatigue was also at cast and wrought levels. Later work with ADMA Products TiH_2 powder was equally successful. In other work, angular metalysis powder[14] was first converted to a spherical morphology (though this morphology change can add US$70/lb to the overall cost of the powder) and then fabricated by AM. Such an example is a turbo charger for an auto engine shown in Fig. 7.2. Interestingly, primary

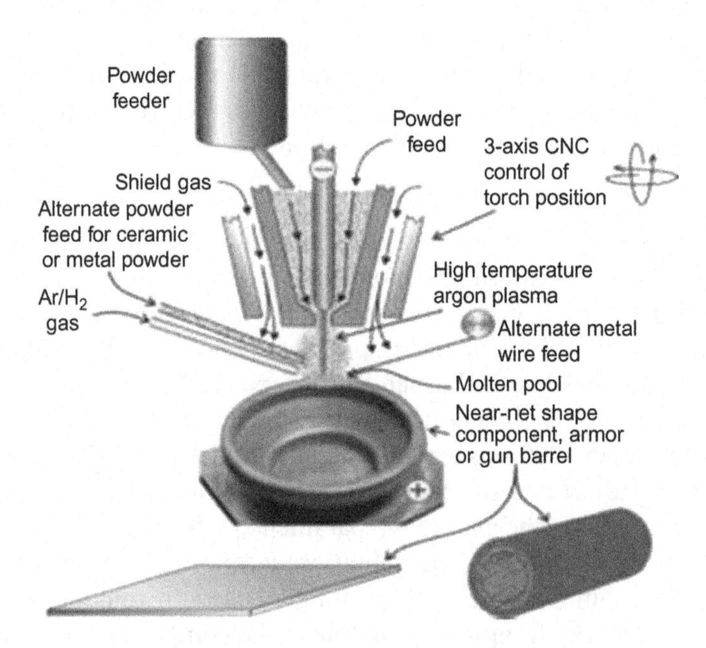

Figure 7.1 Schematic of the MER plasma transferred arc (PTA) additive manufacturing process.[13]

Figure 7.2 An automotive component produced from metalysis powder after spheriodization of the powder.[13]

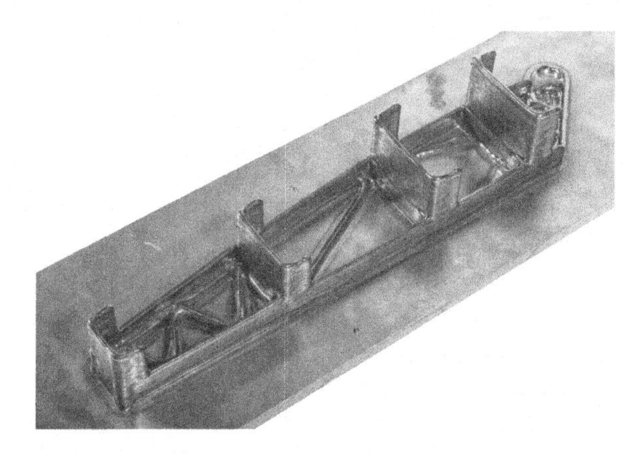

Figure 7.3 Ti–6Al–4V flap rib demonstrator part (9 kg weight) built using WAAM process.[16]

work on low-cost titanium AM has targeted the automotive applications, low cost always being a primary concern with the auto industry.

7.6 LARGE PART CAPABILITIES

Recent interest in titanium AM has focusing on larger aerospace parts using DED technology. Researchers at the State Key Laboratory in China built a Ti–6Al–4V wing spar cap strips for the C919 aircraft using AM.[15] The strip length is ~3100 mm. The high cycle fatigue life of this sample is reported to be higher than the forging part. Similar large titanium demonstrator parts weighing from 9 kg (20 lbs) to 24 kg (53 lbs) have been fabricated by the WAAM process at Cranfield University, UK, for the aerospace industry (Fig. 7.3).[16] Binder Jetting technology is also being explored to form large titanium parts and

Figure 7.4 Ti–6Al–4V valve body built using binder jetting technology.[17] Source: Courtesy of Puris LLC.

Fig. 7.4 shows one such example of a titanium valve body weighing about 11 kg (25 lbs).[17] One of the challenges in AM of large parts is the associated residual stress and resulting distortion or cracking and other defects generation. Intermittent stress relieving using techniques such as high frequency laser shock peening or ultrasonic impact peening is being explored as potential ways to mitigate the residual stress, while processing large size parts.[18]

7.7 NEW INSPECTION TECHNOLOGIES

Another major research focus for AM is dedicated towards process monitoring, control and in-line inspection. As AM involves layer-by-layer build-up, it allows unique opportunity for in-line inspection of each layer during the build. Various investigations are underway to develop in-line inspection tools for AM processes using ultrasonic, X-ray and other tools. A combination of the latest sensor technologies and predictive control algorithms have allowed achievement of compensation for over or under-build of a layer caused by issues due to toolpath overlap or powder catchment.[19,20] On-line spectroscopic analysis of meltpool plasma has yielded promises of monitoring and possibly controlling chemistry of build material.[18] The following areas of development have been identified as high priority from an inspection and quality control perspective[11]:

- Lack of capability for high-speed video and high-speed thermograph
- Lack of real-time material monitoring and measurements
- Inability to perform in situ detection of processing anomalies leading to discontinuities: thermal gradients, voids, and inclusions
- Ineffective feedback control for material composition and microstructure; sensor integration is unattainable with current black box controllers.

7.8 PROJECTED USE OF AM FOR FABRICATION OF TITANIUM COMPONENTS

Looking into the future is always a difficult endeavor, but peering into a murky crystal ball and being optimists as to the future, our speculations for the future of additive manufacturing are set out below. The time at which the projections were set is mid-2016.

The aerospace industry has long recognized the benefits of high-performance titanium alloys for reducing weight in fuselage structures and aircraft skins, interior appliances, aero engine components, and aircraft landing gears. Sophisticated alloys are found in engine fan blades and discs, front bearing housings, compressor and turbine blades, discs, vanes, and hydraulic tubing. For the aerospace industry (military and commercial) the potential use of the AM technique is particularly attractive because of the high buy-to-fly ratios which are often found with these parts (numbers as high as 20:1 often occurring). The potential use of additive manufacturing of titanium components in this industry is 155 tonnes per year at the 5 year point and 400–800 tonnes per year at the 10 year point.[21,22]

Titanium alloys are being evaluated for use in artificial hips and knees and osteosynthesis (surgical procedures that stabilize and repair broken bones by using mechanical devices such as metal plates, pins, rods, wires or screws). For the medical industry, the potential use of additive manufacturing of titanium components is 275 tonnes per annum 5 years from now and 600–1000 tonnes per year 10 years from now.[21,22]

Table 7.1 Some of the Applications for Titanium Components Being Used in Automobiles[23]

Component	Material	Manufacturer	Model
Connecting rods	Ti–3Al–2V–rare earth	Honda	Acura NSX
Connecting rods	Ti–6Al–4V	Ferrari	All 12-cyl.
Wheel rim screws	Ti–6Al–4V	Porsche	Sport wheel option
Brake pad guide pins	Ti grade 2	Daimler	S-Class
Brake sealing washers	Ti grade 1 s	Volkswagen	All
Gearshift knob	Ti grade 1	Honda	S2000 Roadster
Connecting rods	Ti–6Al–4V	Porsche	GT3
Valves	Ti–6Al–4V and PM–Ti	Toyota	Altezza 6-cyl.
Turbo charger wheel	Ti–6Al–4V	Daimler	Truck diesel
Suspension springs	TIMETAL LCB	Volkswagen	Lupo FSI
Wheel rim screws	Ti–6Al–4V	BMW	M-Techn. Option
Valve spring retainers	β-titanium alloys	Mitsubishi	All 1.8l 4-cyl.
Turbo charger wheel	γ-TiAl	Mitsubishi	Lancer
Exhaust system	Ti grade 2	General Motors	Corvette Z06
Wheel rim screws	Ti–6Al–4V	Volkswagen	Sport package GTI
Valves	Ti–6Al–4V and PM–Ti	Nissan	Infiniti Q45
Suspension springs	TIMETAL LCB	Ferrari	360 Stradale

Note: *Many of these parts are "chunky" in nature and can be therefore potentially fabricated by AM.*

A number of automobile original equipment manufacturers (OEMs) are actively looking at titanium AM for chunky component fabrication. Examples include connecting rods, gearshift knob, valves, turbo charger wheel, suspension springs, exhaust system, and wheel rim screws (Table 7.1).[23] If the low-cost titanium powder manufacturing and high throughput PTA based AM processes become reality,[13] AM can have a breakthrough in this highly cost-sensitive automotive market. Projections are for 500 tonnes of AM parts at the 5 year point and 1000–1500 tonnes at the 10 year point.

The defense sector's continual need to reduce equipment weight has created many opportunities for titanium. From ballistic armor to engine and body components, titanium is a lightweight alternative to steel in a growing number of applications. Low volume production needs with higher per part cost makes defense industry an attractive market for AM application. The potential use of Ti metal powders in additive manufacturing for the defense industry (that is nonaerospace) can potentially be quite significant but is difficult to predict.

There may also be applications for titanium components fabricated by AM in the chemical processing industry and oil and gas exploration/distribution, but it is not considered that these will be very extensive. Various other industries are investigating the possibility of using AM for fabricating titanium components.

Total additive manufacturing components using titanium is potentially 930 tonnes per year at the 5 year point and 2000–3500 tonnes per year at the 10 year point. If AM can break the cost barriers of automotive industry, the titanium usage can be as high as 2000 tonnes in 5 years and 5000–6000 tonnes in 10 years. As the AM technologies continue to mature, the titanium powder prices are expected to lower while process speed gets faster. A combination of these factors will boost titanium additive manufacturing and assist a continuous growth of this exciting technology.[22]

REFERENCES

1. The Economist, February 10, 2011. <http://www.economist.com/node/18114221/print>.
2. Columbus L. Forbes/Tech, March 31, 2015. <http://www.forbes.com/sites/louiscolumbus/2015/03/31/2015-roundup-of-3d-printing-market-forecasts-and-estimates/#746b9401dc67>.
3. Collins PC, Banerjee R, Banerjee S, Fraser HL. Laser deposition of compositionally graded titanium/vanadium and titanium/molybdenum alloys. *Mater Sci Eng* 2003;**A352**:118–28.
4. Lin X, Yue TM, Yang HO, Huang WD. Solidification behavior and the evolution of phase in laser rapid forming of graded Ti6Al4V–Rene88DT Alloy. *Metall Trans A* January 2007;**38A**:127–37.
5. Pang W, Man HC, Yue TM. Laser surface coating of Mo–WC metal matrix composite on Ti6Al4V alloy. *Mater Sci Eng A* 2005;**390**:144–53.
6. Filip R. Alloying of surface layer of the Ti–6Al–4V titanium alloy through the laser treatment. *J. Achieve Mater Manuf Eng* March–April, 2006;**15**(1–2):174–80.
7. Tang HP, Yang GY, Jia WP, He WW, Lu SL, Qian M. *Mater Sci Eng* 2015;**A636**:103–7.
8. Kelly SM, Kampe SL. Microstructural evolution in laser-deposited multilayer Ti–6Al–4V builds. Part II. Thermal modelling. *Metall Mater Trans A* 2004;**35A**:1869–79.
9. Makiewicz K, Babu SS, Keller M, Chaudhary A. Microstructure evolution during laser additive manufacturing of Ti–6Al–4V alloy. In: Trends in welding research 2012: Proceedings of the 9th international conference. ASM International, February 01, 2013, p. 970–7.
10. Dehoff RR, Kirka MM, Sames WJ, Bilheux H, Tremsin AS, Lowe LE, et al. *Mater Sci Tech* 2015;**31**(8):931–8. Available from: http://dx.doi.org/10.1179/1743284714Y.0000000734.
11. Workshop report on "Measurement Science Roadmap for Metal-based Additive Manufacturing". NIST, US Dept. of Commerce; May 2013.
12. Bourell DL, Leu MC, Rosen DW. Roadmap for additive manufacturing identifying the future of freeform processing. The University of Texas at Austin, Laboratory for Freeform Fabrication, Advanced Manufacturing Center; 2009. p. 11–5.

13. Withers JC, Shapovalov V, Storm R, Loutfy RO. There is low cost titanium componentry today, Reference 5, 11.
14. Whittaker P. “Metalysis” titanium powder used to 3D print automotive parts. <http://www.ipmd.net/news/002519.html> [accessed February 2014].
15. Huang W, Lin X. 3D printing; 2014. p. 156–65. Available from: http://dx.doi.org/10.1089/3dp.2014.0016.
16. Addison A, Ding J, Martina F, Lockett H, Williams S, Zhang X. In: Proceedings of Titanium Europe 2015, Birmingham, UK, May 2015.
17. Bono E. Puris LLC. Private Communication; March 2016.
18. Sidhu J, Andrew D, Potter M. Intl. Patent PCT WO2014/072699 A1; May 2014.
19. Dutta B, Palaniswamy S, Choi J, Song LJ, Mazumder J. Additive manufacturing by direct metal deposition. *Adv Mater Process* May 2011;33–6.
20. Song L, Singh-Bagavath V, Dutta B, Mazumder J. Control of melt pool temperature and deposition height during direct metal deposition process. *Int J Adv Manuf Technol* May 2011.
21. Withers JC. MER Corp. Private Communication; May 11, 2016.
22. Market Spotlight. Adv Mater Process 2016;174:6.
23. Froes FH (Sam), et al. Titanium in the family automobile: the cost challenge. In: Froes FH (Sam), Imam MA, Fray D, editors. *Cost affordable titanium symposium.* Warrendale, PA: TMS; 2004. p. 159–66.

Printed in the United States
By Bookmasters